Smart Electrical Grid System

Smart Engineering Systems: Design and Applications

Series Editor:
Suman Lata Tripathi

Internet of Things
Robotic and Drone Technology
Edited by Nitin Goyal, Sharad Sharma, Arun Kumar Rana, and Suman Lata Tripathi

Smart Electrical Grid System
Design Principle, Modernization, and Techniques
Edited by Krishan Arora, Suman Lata Tripathi, and Sanjeevikumar Padmanaban

For more information about this series, please visit: https://www.routledge.com/

Smart Electrical Grid System
Design Principle, Modernization, and Techniques

Edited by
Krishan Arora, Suman Lata Tripathi, and
Sanjeevikumar Padmanaban

CRC Press
Taylor & Francis Group
Boca Raton London New York

CRC Press is an imprint of the
Taylor & Francis Group, an **informa** business

MATLAB® is a trademark of The MathWorks, Inc. and is used with permission. The MathWorks does not warrant the accuracy of the text or exercises in this book. This book's use or discussion of MATLAB® software or related products does not constitute endorsement or sponsorship by The MathWorks of a particular pedagogical approach or particular use of the MATLAB® software

First edition published 2023
by CRC Press
6000 Broken Sound Parkway NW, Suite 300, Boca Raton, FL 33487-2742

and by CRC Press
4 Park Square, Milton Park, Abingdon, Oxon, OX14 4RN

CRC Press is an imprint of Taylor & Francis Group, LLC

ISBN: 978-1-032-15048-2 (hbk)
ISBN: 978-1-032-15049-9 (pbk)
ISBN: 978-1-003-24227-7 (ebk)

DOI: 10.1201/9781003242277

Typeset in Times
by codeMantra

Contents

Preface

A smart electrical grid serves several purposes, and the transition from traditional electric grids to smart grids is driven by multiple factors, including the deregulation of the energy market, evolutions in metering, changes in the level of electricity production, *distributed energy system*, changing regulations, the increase in microgeneration and microgrids, renewable energy sources, and new purposes for which electricity is needed (*e.g., electrical vehicle charging points*). The smart grid enables electricity producers to enhance reliability, availability, and efficiency. It can provide cost savings—both for utilities and for consumers—and even environmental benefits. It will allow for faster restoration of power after outages and supports better integration of distributed generation resources, including renewable energy, to the larger grid. This book also supports the grid integration of large-scale renewable energy resources along with the challenge to lower the losses incurred with efficient power transmission.

Load frequency control approaches also play a significant role in modeling, analysis, prediction of the performance, and effective control and utilization of smart energy systems. This book will cover smart techniques that are fruitful at different levels of design and development of smart electrical grids and next-generation smart energy systems. The smart electrical grid will become a potential area of research when every underdeveloped country is trying to focus on smart cities and urban planning with a new level and standards.

Wind and solar generation methods by their inherent characteristics are variable and pose a severe challenge to grid security and system stability. To facilitate renewable integration into the grid, transmission and balancing reserves are required. In conventional power grids, the outer areas of cities and rural areas were built with technologies that are decades old. These grids were considered the peak of energy engineering. As decades have passed, new advancements in the energy industry have been made, and older power grids are being replaced with smarter and more efficient grids that come with better energy management and information tracking. With access to real-time data, based on household energy consumption and its major contributors, energy companies can assist homeowners to effectively manage their energy usage. Using the data, they can make informed recommendations about behavioral changes that will have a clear impact on a household's energy efficiency.

In short, the digital technology that allows for wireless communication between the utility and its customers, and the sensing along the transmission lines is what makes the grid smart. Like the Internet, the smart grid will consist of controls, computers, automation, and new technologies and equipment working together, but in this case, these technologies will work with the electrical grid to respond digitally to our quickly changing electric demand.

This book presents novel concepts in the development of smart cities and smart grids and discusses the technologies involved in producing efficient and economically feasible energy technologies around the world. Also, the proposed book serves the needs of industry professionals, researchers, and doctoral students studying energy technologies. This book will mainly focus on three areas of electrical engineering that are control systems, power electronics, and renewable resources, including artificial intelligence for the development of the smart electrical grid.

MATLAB® is a registered trademark of The MathWorks, Inc. For product information, please contact:
The MathWorks, Inc.
3 Apple Hill Drive
Natick, MA 01760-2098 USA
Tel: 508-647-7000
Fax: 508-647-7001
E-mail: info@mathworks.com
Web: www.mathworks.com

Editors

Dr. Krishan Arora completed his PhD in the area of Electrical Engineering in MM (Deemed to Be) University, Ambala. He did his MTech in Electrical Engineering in IKG Punjab Technical University, Punjab, and BTech in Electrical and Electronics Engineering in IKG Punjab Technical University, Punjab. He is associated with Lovely Professional University as an Associate Professor with more than 13 years of experience in academics. He has been serving as Head of Department (Labs) in School of Electronics and Electrical Engineering, Lovely Professional University, since February 2017. He has published more than 40 research papers in refereed IEEE, Springer, and IOPscience journals and conferences. He has organized several workshops, summer internships, and expert lectures for students. He has supervised more than 10 postgraduate theses and more than 15 undergraduate student's projects. He has completed 15 non-government and consultancy projects. He has attended/participated in 24 national/international online webinars. His area of expertise includes electrical machines, non-conventional energy sources, load frequency control, automatic generation control, and modernization of smart grids. He has taught various courses to UG and PG levels, such as power electronics, non-conventional energy sources, electric drives, induction and synchronous machines, and digital electronics.

Dr. Suman Lata Tripathi completed her PhD in the area of microelectronics and VLSI in MNNIT, Allahabad. She did her MTech in Electronics Engineering in UP Technical University, Lucknow, and BTech in Electrical Engineering in Purvanchal University, Jaunpur. She is associated with Lovely Professional University as a Professor with more than 17 years of experience in academics. She has published more than 55 research papers in refereed IEEE, Springer, and IOPscience journals and conferences. She has also published 9 Indian patents and 1 copyright. She has organized several workshops, summer internships, and expert lectures for students. She has worked as a session chair, conference steering committee member, editorial board member, and reviewer in international/national IEEE journals and conferences. She received the "Research Excellence Award" in 2019 at Lovely Professional University. She had received the best paper at IEEE ICICS-2018. She has edited more than 12 books/1 book series in different areas of electronics and electrical engineering. She is associated for editing work with top publishers such as Elsevier, CRC Press/Taylor and Francis, Wiley-IEEE, SP Wiley, Nova Science, and Apple Academic Press. She has published the edited books *Recent Advancement in Electronic Device, Circuit and Materials* (Nova Science Publishers, 2020), *Advanced VLSI Design and Testability Issues* (CRC Press/Taylor & Francis, 2022), and *Electronic Devices and Circuit Design Challenges for IoT Application* (Apple Academic Press, 2020). She is also an editor of *Green Energy: Fundamentals, Concepts, and Applications* (Wiley-Scrivener Publishing, 2021) and *Design and Development of Energy Efficient Systems* (Wiley-Scrivener Publishing, in press). She is also associated with Wiley-IEEE for her multi-authored (ongoing) book in the area of VLSI design with HDLs. She is also working as the series editor for the title *Smart Engineering Systems* (CRC Press/Taylor & Francis, 2022). She has already completed one book with Elsevier titled *Electronic Device and Circuits Design Challenges to Implement Biomedical Applications* (2021). She is a guest editor of a special issue of Bentham Science titled *Current Medical Imaging* (2022) She is a senior member of IEEE, Fellow IETE, and life member of ISC and continuously involved in different professional activities along with academic work. Her area of expertise includes microelectronics device modeling and characterization, low-power VLSI circuit design, VLSI design and testing, and advance FET design for IoT, embedded system design, and biomedical applications.

Dr. Sanjeevikumar Padmanaban (Member'12–Senior Member'15, IEEE) received his bachelor's degree in electrical engineering from the University of Madras, Chennai, India, in 2002, his master's degree (Hons.) in electrical engineering from Pondicherry University, Puducherry, India, in 2006, and his PhD degree in electrical engineering from the University of Bologna, Bologna, Italy, in 2012. He worked as an Associate Professor in VIT University from 2012 to 2013. In 2013, he joined the National Institute of Technology, India, as a Faculty Member. In 2014, he was invited as a Visiting Researcher by the Department of Electrical Engineering, Qatar University, Doha, Qatar, funded by the Qatar National Research Foundation (Government of Qatar). He continued his research activities with the Dublin Institute of Technology, Dublin, Ireland, in 2014. Further, he served an Associate Professor with the Department of Electrical and Electronics Engineering, University of Johannesburg, Johannesburg, South Africa, from 2016 to 2018. Since 2018, he has been a Faculty Member with the Department of Energy Technology, Aalborg University, Esbjerg, Denmark. He has authored over 300 scientific papers. Dr. S. Padmanaban was the recipient of the Best Paper *cum* Most Excellence Research Paper Award from IET-SEISCON'13, IET-CEAT'16, IEEE-EECSI'19, and IEEE-CENCON'19 and five best paper awards from ETAEERE'16 sponsored Lecture Notes in Electrical Engineering, Springer book. He is a Fellow of the Institution of Engineers, India, the Institution of Electronics and Telecommunication Engineers, India, and the Institution of Engineering and Technology, the UK. He is an Editor/ Associate Editor/Editorial Board for refereed journals, in particular the IEEE Systems Journal, IEEE Transaction on Industry Applications, IEEE Access, *IET Power Electronics, IET Electronics Letters,* and *International Transactions on Electrical Energy Systems*, Wiley; a Subject Editorial Board Member for *Energy Sources—Energies Journal*, MDPI; and the Subject Editor for the *IET Renewable Power Generation, IET Generation, Transmission and Distribution*, and *FACTS* journal (Canada).

Contributors

Ashu Ahuja
Electrical Engineering Department
M. M. University
Mullana, India

Krishan Arora
School of Electronics and Electrical
    Engineering
Lovely Professional University
Phagwara, India

T. Chinnadurai
Department of Mechatronics
Reva University
Bangalore, India

Shakti Raj Chopra
School of Electronics and Electrical
    Engineering
Lovely Professional University
Phagwara, India

Vineet Dahiya
Department of Electrical & Electronics
    Engineering
School of Engineering & Technology
K. R. Mangalam University
Gurgaon, India

Radhika G. Deshmukh
Department of Physics
Shri Shivaji Science College
Amravati, India

Dhanaselvam J.
Department of Instrumentation and
    Control Engineering
Sri Krishna College of Technology
Coimbatore, India

Amit Kumar Dhir
Department of Civil Engineering
Lovely Professional University
Phagwara, India

Aman Ganesh
School of Electronics and Electrical
    Engineering
Lovely Professional University
Phagwara, India

A. Gayathri
Department of EEE
Sri Krishna College of Technology
Coimbatore, India

A. Gomathi
Department of Energy Science and
    Technology
Periyar University
Salem, India

Kapil Kumar Goyal
Dr. B. R. Ambedkar National Institute
    of Technology
Jalandhar, India

Vikram Kumar Kamboj
School of Electronics and Electrical
    Engineering
Lovely Professional University
Phagwara, India
and
Department of Electrical and
    Computer Engineering,
    Schulich School of
    Engineering
University of Calgary
Alberta, Canada

M. Karthigai Pandian
Department of ECE
GITAM School of Technology
Bangalore, India

Asif Iqbal Kawoosa
Department of Computer Applications
Lovely Professional University
Phagwara, India

Gaurav Kumar
Vidya College of Engineering
Meerut, India

Mukesh Kumar
Vidya College of Engineering
Meerut, India

Pawan Kumar
Department of Computer Science and
Application
NIILM University
Kaithal, India

P. Maadeswaran
Department of Energy Science and
Technology
Periyar University
Salem, India

V. Manimegalai
Department of EEE
Sri Krishna College of Technology
Coimbatore, India

Tanuj Mishra
School of Electronics and Electrical
Engineering
Lovely Professional University
Phagwara, India

P. Pandiyan
Department of EEE
KPR Institute of Engineering and
Technology
Coimbatore, India

T. Prabhuraj
Department of Energy Science and
Technology
Periyar University
Salem, India

Deepak Prashar
School of Computer Science and
Engineering
Lovely Professional University
Phagwara, India

K. A. Ramesh Kumar
Department of Energy Science and
Technology
Periyar University
Salem, India

Umesh C. Rathore
Electrical Engineering
Department
Govt. Hydro Engineering College
Bandla (Bilaspur), India

V. Rukkumani
Department of Electronics and
Instrumentation
Sri Ramakrishna Engineering College
Coimbatore, India

K. Saravanakumar
Department of Instrumentation and
Control Engineering
Sri Krishna College of Technology
Coimbatore, India

M. P. Sharma
Planning Department
Rajasthan Rajya Vidyut Prasaran Nigam
Limited (RVPN)
Jaipur, India

Pushpendra Kumar Sharma
Departments of Civil Engineering
Lovely Professional University
Phagwara, India

Shelja
Department of Computer Application
Tilak Raj Chadha Institute of
    Management and Technology
    (TIMT)
Yamunanagar, India
and
Department of Computer Science and
    Application
NIILM University
Kaithal, India

Amit Kumar Singh
School of Electronics and Electrical
    Engineering
Lovely Professional University
Phagwara, India

Sanjeev Singh
Electrical Engineering Department
Maulana Azad National Institute of
    Technology (MANIT)
Bhopal, India

Bhawna Tandon
Electronics & Communication
    Engineering Department
Chandigarh Engineering College
Landran, India

Pradeep Singh Thakur
Electrical Engineering Department
RG Govt. Engineering College
Kangra (Nagrota Bagwan), India

Abhishek Tyagi
Department of Mechanical Engineering
Dr A P J Abdul Kalam Technical
    University
Lucknow, India

Bhavesh Vyas
Department of Electrical & Electronics
    Engineering
School of Engineering & Technology
K. R. Mangalam University
Gurgaon, India

1 Internet of Things-Based Modernization of Smart Electrical Grid

Krishan Arora
Lovely Professional University

CONTENTS

1.1 INTRODUCTION

Automation is evolving in every field around us. As a person living in the fastest growing age of technology development, we have moved from room-sized computers to light and portable laptops and from insecure databases to highly secure web browsers. The same is the case for electrical grids. There are several technological developments in this field of automation, and one of them is automation of grid or "smart electrical grid" (SEG). By connecting individual devices and various appliances to a centralized hub and controlling each and every thing through that network is quite fascinating. An SEG allows you to experience high-tech functionality and convenience that wasn't possible in the past.

IoT has numerous applications in different fields, such as e-health, smart home, smart cities, smart education system, smart transportation, smart agriculture, and smart factories. Above all, SEG has major attraction in academia and industry due to its direct relation with the real-time events [1–3]. Figure 1.1 depicts the various applications of the smart grid [4,5].

DOI: 10.1201/9781003242277-1

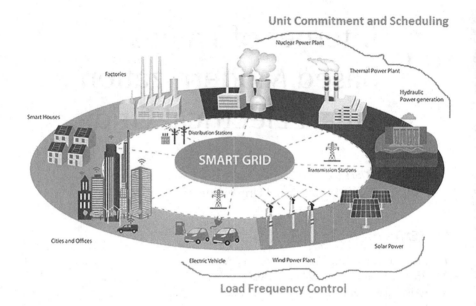

FIGURE 1.1   Modern smart grid system.

SEGs can be classified into two broad categories based on controlling methodology: (i) native controlled and (ii) remotely controlled [6]. The former system depicts that users can control grid appliances by using controllers via wireless technologies (e.g., Zigbee, GSM, and Bluetooth) or stationary communication methods, whereas in the case of latter, users can control the electrical grid with Internet connections through mobile phones. Numerous flaws occur while setting up the controlling methodology; thus, the SEG should provide an interface for observing and controlling the appliances [6].

1.2   PROS OF SMART ELECTRICAL GRID (SEG)

Nowadays, for a fast and reliable connection, wireless communication systems such as Wi-Fi are more promising candidates with ample advantages over wired connections:

   i. **Cost-Effective**: The installation cost decreases due to the non-requirement of cables in contrast to the wired system where the cable, its material, and laying technology along with the need for experienced professionals add up to the cost.
  ii. **Easy Deployment, Installation, and Coverage**: Remote devices or appliances can be easily accessed wirelessly along with easy deployment and installation.
 iii. **System Extension and Scalability**: Transportation of a native structure is predominantly priceless when, due to novel or distorted prerequisites, augmentation of the scheme is essential. More than wired establishments, in

which cabling extension is expensive and time-consuming [7], this makes remote systems an original essence.

iv. **Amalgamation of Mobile Devices**: With the advancement of the communication technology, wireless platform gives the automation an easier and possible solution, where devices or appliances can be remotely controlled or observed without any physical connections.

v. **Aesthetic Benefits**: This not only covers a large area, but also beautifully manages the architecture and gives immense benefits to historical buildings and tourist spots.

1.3 EXISTING SYSTEM

Currently, modern electrical grids use a lot of techniques. Different modes of wireless communications are used to transfer data and signals between their various components, such as Bluetooth, Zigbee, Wi-Fi, and Global System for Mobile Communications (GSM) [8]. Bluetooth (HC-05) acts as a slave device and is used for communication between Arduino (Mega) IDE and a smartphone. The operating voltage is 3.6–6V. The Bluetooth Board HC-06 has 6 pins such as State, RXD, TXD, GND, VCC, and EN. Internally, the input/output port of an Arduino board (microcontroller) is programmed using C language/Python and the connection is made to home appliances through Bluetooth BT cell [9].

Zigbee is under radio frequency (RF) communication and uses IEEE 802.15.4 standard. The reflection of the Zigbee standard illustrates the packet routing in a network where a large number of nodes exist and the transmission of data occurs in a zigzag pattern as shown in Figure 1.2. Zigbee supports application in all fields. The main component in Zigbee is its coordinator (ZC), which maintains the network and routing table. This standard has three frequency bands, and the prominent band of use is 2.5 GHz with data rates of 250 and 40 kbps. This method has numerous

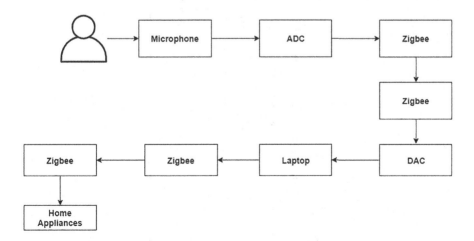

FIGURE 1.2 Zigbee-based automation system.

TABLE 1.1

Comparison between Different Methodologies

Methodologies	Communication Interface	Controller	User Interface	No. of Devices	Cost	Speed	Range	Energy Savings
Bluetooth	Bluetooth and AT instructions	Arduino Uno	Android	Unlimited	Moderate	Low	10m	No
GSM	SMS messages	Arduino ADK	Mobile phone	Limited	Low	Very low		No
Zigbee	Zigbee with instructions	Zigbee controller (ZC)	Smartphone	Unlimited	High	Low	10m	Yes
The proposed wireless design	Wi-Fi (NODEMCU)	Arduino SDK	Smartphone	Unlimited	Low	High	Unlimited	Yes

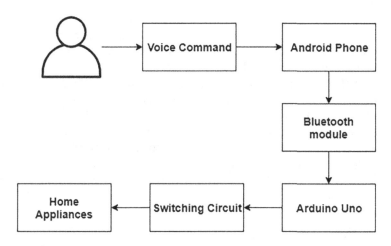

FIGURE 1.3   Voice recognition.

advantages as per the literature, such as low cost, interoperability, low signal-to-noise ratio, low power consumption, implementation by using any microcontroller, easy deployment of innovations, high data security, and requirement of low bandwidth for minor projects [10] (Table 1.1).

As per the literature, voice control method is used for SEG as shown in Figure 1.3 with the help of Arduino and various wireless technologies such as GSM, Bluetooth, Wi-Fi and NFC. In the reported literature [11–13], the experiment is conducted on five groups with different voice commands to access fan, light, and utility, and the safety check was performed at a baud rate of 115,200 bps to make an SEG [14]. In addition to smart grid automation, the advantage of the feature along with voice is that the motion of the objects or appliances can be recognized [15]. The grid automation is controlled through voice by using Arduino as a controller, Bluetooth HC-05 module as a wireless communicator, and HC-SR04 as an ultrasonic sensor to detect the motion as shown in Figure 1.3.

1.4   RESEARCH GAPS

Technology is advancing in every field, and that's how the concept of grid automation emerged. Technology is all about making each and every task easier. There are different technologies that are used to automate electrical grids which are discussed. The development and implementation has many problems as follows:

    i. **Dependency on Range**: The SEG connects all the devices of the grid to the mobile app through Bluetooth. It works when a person is in the range of the central hub. Generally, the range of Bluetooth connectivity is 100 m [16]. Your grid appliances cannot be controlled outside of this range.

   ii. **Expandability**: New grid assistants such as Alexa and Google Assistant are trending a lot these days, but the appliances they interact with are limited. They can only interact with smart appliances and won't control our standard grid appliances such as transformers and insulators [16] [17]. The other way to control them is hardwiring, which again is an expensive alternative.

iii. **Economical**: The main motive of technology is to empower each and every individual, but sometimes the cost of new technology is a barrier [18]. Being new in the market, grid automations are quite expensive.

iv. **Energy Saving**: In this regard, electricity can be saved by automatically switching off the appliances when not in use. Similarly, nowadays, the smart grid system [19] helps save energy with the implementation of smart condenser banks which cover all reactive power generated by the current transformers (CTs) and potential transformers (PTs). Monitoring the electricity usage can be helpful in saving the energy and predicting the cost for an individual.

1.5 PROPOSED SYSTEM

In the proposed system, there are three modules in total. Each module is made for a different purpose, and each of them functions differently. The mobile application is to provide a user-friendly interface that will help organize and observe the usage of grid appliances. For this module, the enhanced features of Android and Java programming language are used. The purpose of the second module is to connect the application to a database; this can be achieved through Firebase. The third module in the system is to connect devices to a central hub, which then sends and receives data from server. This feature is enabled using IoT devices and C programming language.

1.6 METHODOLOGY

1.6.1 FRAMEWORK

In Figure 1.4, the design flow is shown; after analyzing the functionality and feasibility of the problem, the requirements of hardware and software for communication are explored in the following section. The software section consists of the mobile application design with log-in and registration page along with a home page and a dedicated page for each device. In addition, more functionalities to the application, such as monitoring energy usage and controlling devices, are added as shown in

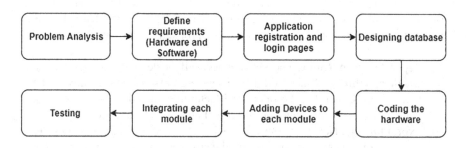

FIGURE 1.4 Design flow of the propo.sed system.

Figure 1.4. Then the designing of the real-time database to store the state of each device in the system is performed. Code is written to connect the hardware components to the database created earlier along with the application. After this, an authentication and authorization page is created that enables log in through Google and connects different home appliances to our central hub, such as fan, bulb, and mobile charger. Finally, the JUnit testing is done for modules. The manual application testing on different Android devices running different versions from Lollipop to Android 10 is explored [19].

1.6.2 COMPONENT REQUIREMENT ANALYSIS

The requirements of the projected system can be majorly categorized into hardware and software. Thus, in this segment, the analysis of the hardware components and software is carried out. The features and demanded configuration of the microcontroller are studied along with compatible software packages.

a. **Hardware Requirements**
   I. Relay (an electrically operated switch).
   II. Wi-Fi module (ESP8266 NODEMCU) (integrated TR switch, low-noise amplifier, TCP/IP protocol, integrated PLL, and regulator and power organization section, along with 19.5 dBm output power in 802.11b mode as shown in Figure 11.5).
   III. Jumper wires.
   IV. General-purpose PCB.
   V. Temperature and humidity sensor (DHT11) (digital-type capacitive humidity sensor; it uses thermistor for measuring the outside air and sends digital signal on data pins).

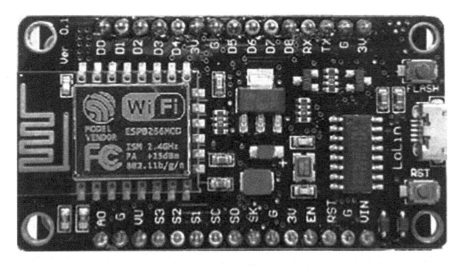

FIGURE 1.5   Wi-Fi module (ESP8266 NODEMCU).

b. **Software Requirements**
    I.   Android Studio 3.0.1.
    II. Python 2.7.
   III. Arduino IDE 1.8.9.

1.7 IMPLEMENTATION

The flow diagram shown in Figure 1.6 depicts that the user can control and observe the functionalities of the developed application that incorporates the devices (visualization, addition, deleting, and controlling), along with the energy consumption reading feature, which has been newly added to the SEG. Moreover, the maintenance or support feature is controlled by the administrator in addition to customers.

    All related information is shared in the cloud database that runs on cloud computing platform, where the data can be accessed using software as a service. In database services, scalability and high availability of the database are ensured by the database service provider. As per the design, the Firebase platform provided by Google offers various services such as storage, real-time database, online processing, and authorization of the user along with their updated usage information, as shown in Figure 1.7.

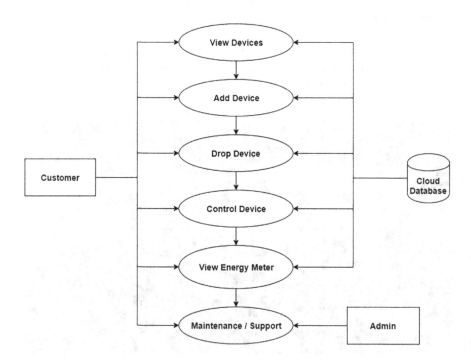

FIGURE 1.6 Functionalities in SEG application.

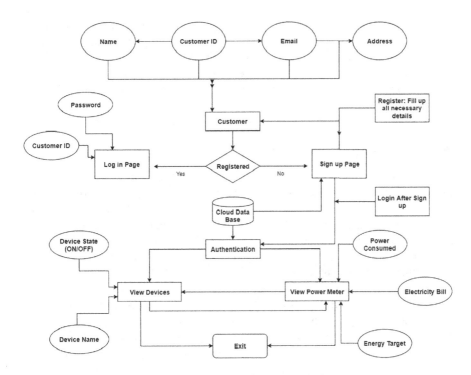

FIGURE 1.7 Implementation of the proposed architecture of application.

1.8 FUNCTIONAL REQUIREMENTS

- **Functional Requirements 1**: Measurement of Parameters: A considerable number of parameters are estimated; for example, current stream, vitality utilization, and time interim for which the gadget is being utilized in a room are important to spare and control the gadgets by utilizing different sensors and versatile applications.
- **Functional Requirements 2**: Fetching of Data: Sensors measure parameters, and those parameters are used for further processes such as data analysis. Measuring and fetching data are the basic functional requirements of this system.
- **Functional Requirements 3**: Manipulation of Data: After fetching the data, manipulation is done with the provided database and the current data. They are analyzed to generate the ideal solution.
- **Functional Requirements 4**: Device Selection: Device selection process looks for the time interval and gives a notification which is used to turn off the device.
- **Functional Requirements 5**: Notification on Mobile Phone: While detecting if there is any device being used for more than its usual time, it notifies the user that something is incorrect.

- **Functional Requirements 6**: Registration/Log-in: Users can log in into the system. They can get the usage time via a notification in the mobile phone, can give feedback for any sort of improvement, and can complain about the error. Admin can manage the entire database and the system and manage the complaints received before and can give best suggestions to them.

1.9   RESULTS AND DISCUSSION

In this section, the consequences based on the accomplishment of the application as proposed are shown using various images, such as the log-in page, registration page, home page, and control page. The log-in page shown in Figure 1.8 provides the option for the user to log in with their credentials into the app. First-time users must register on the registration page in the app as shown in Figure 1.9. In this page, a user can set up their profile by entering e-mail ID and password for further usage of the app.

Also, they'll have to confirm their mail ID by clicking on the link sent to their e-mail ID. The purpose of the e-mail ID is that if a user forgets the password, then it can be retrieved using the e-mail ID, as shown in Figure 1.10. Appliances can be switched on/off using the home page. After logging in to the application, there will be a list of devices available. Devices can be renamed according to the appliances connected to them. The power consumption of each appliance is shown on the home page (see Figure 1.11).

FIGURE 1.8   Log-in page.

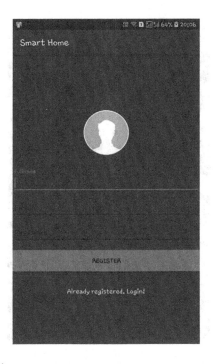

FIGURE 1.9   Registration page.

FIGURE 1.10   Home page.

FIGURE 1.11   Control page.

1.10 CONCLUSIONS

In a nutshell, the implemented app is working successfully with Arduino controller as hardware. The advanced feature is that energy consumption reading is shown in the app; thus, the user can control the saving as per the requirement, along with the addition and deletion of the devices. This chapter also supports the real-time database, online processing, and authentication and authorization of the user by means of the implemented application named as "smart electrical grid."

REFERENCES

1. W. A. Jabbar, M. Ismail, and R. Nordin, "Evaluation of energy consumption in multipath OLSR routing in Smart City applications," in *2013 IEEE 11th Malaysia International Conference on Communications (MICC)*, Kuala Lumpur, Malaysia, 2013, pp. 401–406.
2. K.-M. Lee, W.-G. Teng, and T.-W. Hou, "Point-n-Press: An intelligent universal remote control system for home appliances," *IEEE Transactions on Automation Science and Engineering*, vol. 13, pp. 1308–1317, 2016.
3. P. P. Gaikwad, J. P. Gabhane, and S. S. Golait, "A survey based on Smart Homes system using Internet-of-Things," in *2015 International Conference on Computation of Power, Energy, Information and Communication (ICCPEIC)*, Melmaruvathur, India, 2015, pp. 0330–0335.
4. D. Prashar and K. Arora, Design of two area load frequency control power system under unilateral contract with the help of conventional controller. *IJICTDC*, vol. 5, pp. 22–27, 2020.

5. T. Song, R. Li, B. Mei, J. Yu, X. Xing, and X. Cheng, "A privacy preserving communication protocol for IoT applications in smart homes," *IEEE Internet of Things Journal*, vol. 4, pp. 1844–1852, 2017.

6. K. Patel, K. Arora, and P. Kaur, "Power monitoring system in solar power plant using LabVIEW," in *2019 2nd International Conference on Intelligent Computing, Instrumentation and Control Technologies (ICICICT)*, Kannur, India, 2019, pp. 1011–1015, doi:10.1109/ICICICT46008.2019.8993249.

7. W. A. Jabbar, M. Ismail, R. Nordin, and S. Arif, "Power-efficient routing schemes for MANETs: A survey and open issues," *Wireless Networks*, vol. 23, pp. 1917–1952, 2017.

8. M. N. Jivani, "GSM based home automation system using app-inventor for android mobile phone," *International Journal of Advanced Research in Electrical, Electronics and Instrumentation Engineering*, vol. 3, pp. 12121–12128, 2014.

9. E. M. Wong, "A phone-based remote controller for home and office automation," *IEEE Transactions on Consumer Electronics*, vol. 40, pp. 28–34, 1994.

10. K. Gill, S.-H. Yang, F. Yao, and X. Lu, "A zigbee-based home automation system," *IEEE Transactions on Consumer Electronics*, vol. 55, pp. 422–430, 2009.

11. M. R. Kamarudin, M. Yusof, and H. T. Jaya, "Low cost smart home automation via Microsoft speech recognition," *International Journal of Engineering & Computer Science*, vol. 13, pp. 6–11, 2013.

12. B. Yuksekkaya, A. A. Kayalar, M. B. Tosun, M. K. Ozcan, and A. Z. Alkar, "A GSM, internet and speech controlled wireless interactive home automation system," *IEEE Transactions on Consumer Electronics*, vol. 52, pp. 837–843, 2006.

13. S. Kumar and S. R. Lee, "Android based smart home system with control via Bluetooth and internet connectivity," in *The 18th IEEE International Symposium on Consumer Electronics (ISCE 2014)*, Jeju, Korea (South), 2014, pp. 1–2.

14. Y. Mittal, P. Toshniwal, S. Sharma, D. Singhal, R. Gupta, and V. K. Mittal, "A voice-controlled multi-functional Smart Home Automation System," in *2015 Annual IEEE India Conference (INDICON)*, New Delhi, India, 2015, pp. 1–6.

15. M. E. Abidi, A. L. Asnawi, N. F. Azmin, A. Jusoh, S. N. Ibrahim, H. A. M. Ramli, et al., "Development of voice control and home security for smart home automation," in *2018 7th International Conference on Computer and Communication Engineering (ICCCE)*, Kuala Lumpur, Malaysia, 2018, pp. 1–6.

16. A. Z. Alkar and U. Buhur, "An Internet based wireless home automation system for multifunctional devices," *IEEE Transactions on Consumer Electronics*, vol. 51, pp. 1169–1174, 2005.

17. N. Dickey, D. Banks, and S. Sukittanon, "Home automation using Cloud Network and mobile devices," in *2012 Proceedings of IEEE Southeastcon*, Orlando, FL, USA, 2012, pp. 1–4.

18. Y. Cui, M. Kim, Y. Gu, J. J. Jung, and H. Lee, "Home appliance management system for monitoring digitized devices using cloud computing technology in ubiquitous sensor network environment," *International Journal of Distributed Sensor Networks*, vol. 10, p. 174097, 2014.

19. K. Arora, A. Kumar, V. K. Kamboj, D. Prashar, S. Jha, B. Shrestha, and G. P. Joshi, Optimization methodologies and testing on standard benchmark functions of load frequency control for interconnected multi area power system in smart grids. *Mathematics*, 8, 980, 2020.

2 Role of Non-Conventional Energy Resources in Today's Environment

Pradeep Singh Thakur
RG Govt. Engineering College

Krishan Arora
Lovely Professional University

Umesh C. Rathore
Govt. Hydro Engineering College

CONTENTS

DOI: 10.1201/9781003242277-2

2.1 INTRODUCTION

In the latest years, with the development of living standards and increasing population of the world, the demand for energy is increasing at a fast rate. The developing
countries, with a view to account for 90% of its power demand, call for an increase
in production by 2035. Achieving a greater profitability along with meeting the rising
demand of people is an advanced challenge. Energy safety study come to be greater
customers calls for greater power resources. Such problems of increasing energy
demand can create lots of problems in the upcoming time. It also creates a lots of
opportunities also. We meet this increasing energy demand by burning fossil fuels,
and these fossil fuels release greenhouse gases, specifically carbon dioxide (CO_2),
which is a reason for global warming. These fossil fuels are also decreasing day by
day, so at one-time several human beings lacking approach to electricity keep unac-
ceptably sky-high. There are a number of challenges in the power system, and these
can create many chances for each and every challenge. A sustainable power source is
the need of the hour, and we need to make necessary modifications in the infrastructure to distribute the energy generated without dissipation.

2.2 TYPES OF POWER PLANTS

As the demand for power grows, the community relies increasingly upon fossil fuels
such as coal, oil, and gas. Various conventional resources are used by the people
worldwide in the last two centuries [1]. Main conventional resources used by people
are coal, water, nuclear, and diesel/petrol power plants. We are using these resources
continuously, and these resources are depleting every day. There is a need to find a
sustainable power supply for use in the upcoming days because the cost of gas and oil
goes on increasing every working day. So we want to use increasingly more renewable
resources of power. For the powerful manipulation of renewable resources of power,
there was an established order of an autonomous agency named the *Department
of Non-conventional Energy Sources* by the GoI [2]. Fossil fuels are the biggest
pollutants of the environment. So, due to this demand for renewable power, assets
are growing by every day. These energy resources eliminate the greenhouse consequences and hence also reduce the dependence on oil and gas. Consequently, in order
to fulfill the growing need of the rising community, the technologists/analysts are
trying to develop new strategies for us to faucet into numerous non-traditional assets
of power which are not only of high quality and renewable, but also energy-efficient.
Now we are heading toward the non-traditional resources of energy. The various
non-traditional resources of energy that are used by us nowadays include wind, solar,
biomass, tidal, and geothermal energy. The main conventional energy resources and
non-conventional energy resources are explained below.

2.2.1 NUCLEAR POWER PLANTS

The utilization of a nuclear separation reaction and uranium as material for this
nuclear power plant, a nuclear power plant (PP) generate an excessive quantity of
strength. As nuclear PPs are considered for the supply of low-carbon strength, this

generation is extensively belief of an additional atmospheric conditions for friendly choice [3]. In comparison with renewable resources of power, such as solar and wind, power generation from nuclear PPs is supposed to be more trustworthy. In spite of the reality that the funding necessary to provide power from a nuclear PP is remarkable, the expenditure perturbed into the running of them are correspondingly very low. Nuclear power resources uniformly have a very high mass per unit volume compared to conventional fuels, and they liberate large quantities of power. Because of all these factors, nuclear PPs need a very small amount of nuclear fuel, which can generate a large quantity of power, preparing them especially well organized once they start running up.

2.2.2 HYDROELECTRIC POWER PLANTS

In hydroelectric PPs (HEPPs), hydroelectric power is generated by means of force, i.e., gravitational force of moving or streaming water. Different from other conventional fuel-powered PPs, these PPs release very small amount of CFC and carbon. Nonetheless, the construction of a HEPP and barrage for water storage requires very large amount of funds. In contrast with the global Hydro Electric Power association, 2017 hydro potential, an envisage of around 32 GW of hydro strength potential set off into working conditions in 2016, guide the global accumulative set up potentiality to around 1,250 GW.

2.2.3 COAL-FIRED POWER PLANTS

According to the international association of coal, charcoal-fired PPs accounted for approximately 37% of the global power production in 2018. Charcoal-fired PPs make use of moist coal as a fuel material and produce power and consequently discharge a giant quantity of risky gases into the neighboring environment. In an attempt to lessen the CFC and carbon discharges, many developed nations have previously delivered proposals to eliminate the complications related to charcoal-fired PPs. Canada is trying to make an arrangement to segregate out its charcoal PPs by the end of 2030, while the UK has proposed a deadline of 2025. Other countries such as Germany are focusing toward moving away with the technology from its grid through 2038. Various European global nations are continuing to observe this quickly.

2.2.4 DIESEL-FIRED POWER PLANTS

In these power plants, diesel is used as the fuel. This variety of PP is employed for small-scale production of electricity. These can be set up in places where there is no smooth accessibility to backup electricity resources, and they are used as a support for undisturbed supply of electrical power each and every time there is a breakdown in the electrical supply. Diesel-powered PPs essentially require a much little space and offer better thermal performance in comparison with charcoal-powered PPs. Due to very high level of preservation and expenditure on diesel, diesel PPs have until no longer obtained reputation at identical rate as one of a kind sorts of strength generation PP which incorporates vapor and hydroelectric PP.

2.3   NON-CONVENTIONAL ENERGY RESOURCES

With the exception of by employing the strength of the solar plant to warm the water, dwelling areas or for the generation of power by using the PV cells, we are able to moreover make use of the solar within side the shape of WE to supply electrical strength as it is the sun's solar strength that controls our climate and atmospheric pressure. Our planet is intermittently heated by the radiations emitted by the sun, which makes the air around the equator humid because it absorbs more and more energy and the air near the poles is freezing. When the air in a particular place is heated, it expands and moves upward and it contracts when cooled. This heterogeneity in temperature gives rise to convection currents, which start to move everywhere in the globe as the air which is warm and lighter starts to flow from the colder zones to the hot zones where the air is very much light. This movement of air in the earth's ecosystem from a damp region to a colder place is called the power from the wind and can be robust or vulnerable relying upon the sun's electricity hanging the globe at that moment. In addition, because the globe's land mass and its oceans devour and unencumber sun strength lower back consume and liberate solar energy back into the aerosphere at varied rates, there is a non-stop motion of air around the surface of the globe and the aerosphere offers an upward thrust to the launch in currents, one in producing WE [4]. The earth's spinning also has an essential role in the production of wind strength. We can describe wind as the movement of air, which differs from zero, i.e., the minimum speed to high, i.e., the maximum up to storms. In science, the globe has a limitless delivery of unfastened wind electricity as each zone of the globe gathers the consequences of wind during the daytime for small duration. Furthermore, due to yearly deviations, together with iciness or summer season or weather or geological situations, in conjunction with peak stretches, a little constituents of the area get a supplementary wind strength than others. With the depletion of fossil gasoline substances, WE and wind strength are becoming a vital non-commercial electricity asset in a short time. We appreciate that the wind power is an unfastened and non-commercial secondary form of solar strength due to the uneven allocation of thermal reading in absolute zones throughout the globe, and people have been tackling this unfastened wind strength for their benefit because of the reality that wind generator and cruising boats had been first utilized in historical times. Wind generators exploit the strength in the moving air to convert the mechanical energy into a form of rotating force. This is then employed straightway for water pumping or pounding of corn, although wind generators can also be modified to produce electricity for warming and lighting by coupling a generator to the revolving shank connected to the sails of the windmills. The kinetic strength (kinetic strength is the movement or motion of substances and things) accommodated in the wind most probably can be converted into mechanical and electrical strength via windmill. A modern wind generator that uses the kinetic strength of the wind to deliver electrical energy is called a WT (wind turbine). WMs (windmills) that are in use these days are most probably a type of WG (wind generator) that may be employed in an exclusive manner and in better prosperous way compared to a conventional sail WM. Numerous WMs that are grouped together to grasp massive portions of wind strength and transform it into electrical strength reinforcing this energy into the electrical power grid are referred to as WFs (wind farms). These WFs can be positioned on plain grassland, on top of any peak, or in seaside regions. WT technology can look smooth, but there are various mechanical factors in a modern WT. The wind alternates the rotating shanks of a WT circularly around a critical hub, which revolves the shaft that is attached with a low-pace gearbox, which in turn rotates an electrical generator at a higher velocity and generates strength. This electric generator transforms the kinetic strength of the revolving wings into electrical strength. Electrical wires carry this electrical energy to an electrical SS (substation) for distribution to the electrical grid. Present-day WMs have various air wing-shaped rotor wings harking back to airfoil aircraft propellers, unlike WMs which usually had several flat wings or sails. Even though there are various arrangements of WMs available these days, most of them can be designated as either VAWT (vertical-axis windmill), which have wings that rotate about the axis perpendicular to the direction of the wind, or HAWT (horizontal-axis windmill), which have wings that rotate about the axis parallel to the direction of the wind. The duo has their true and awful factors in which way they try to squeeze the WE (wind energy), but every construction can generate electrical strength that is from around 100 W (one hundred watts) to several hundred watts. Every type incorporates the same identical essential components as follows:

i. A tower and its supporting structures that support the gearbox, generator, rotors, and the auxiliary system.
ii. The rotor blades of the WT, which in particular arrests a large amount of WE.
iii. To give rise to the rotational speed of generator, a mechanical gearbox is attached.
iv. Finally, the electrical energy is generated by an electrical generator.
v. To modify the output and velocity of WM, control equipment and speed sensors are employed with display.
vi. To connect the WT with the electrical grid, electrical cables are used.

2.4   HORIZONTAL- AND VERTICAL-AXIS WT DESIGNS

The majority of the wind power industry is commanded by HAWTs (horizontal-axis wind turbines). The rotating axis of the WT is in the horizontal direction or in alignment with the floor of the HAWT. With enormous wind speeds, HAWTs are used by maximum industries [5]. However, in the areas where the wind speed is low and in domestic applications, VAWTs are used. The advantage of HAWT is all that it can supply appreciable strength from a particular amount of wind. The downside of HAWT, however, is that it's usually bulky and, during stormy conditions, doesn't produce power properly. On the other hand, in VAWTs, the axis of rotation of the WT is in vertical direction or upright direction to the earth. As expressed earlier, VAWTs are mainly used in very small wind plants and in domestic uses. VAWTs utilize wind approaching from all sides. By the reason of this type of adaptability, VAWTs are perceived to have high-quality operation in places where the wind conditions are not compatible or, due to communal mandate, the WT cannot be positioned immoderate enough to gain energy from the regular wind (Figures 2.1 and 2.2).

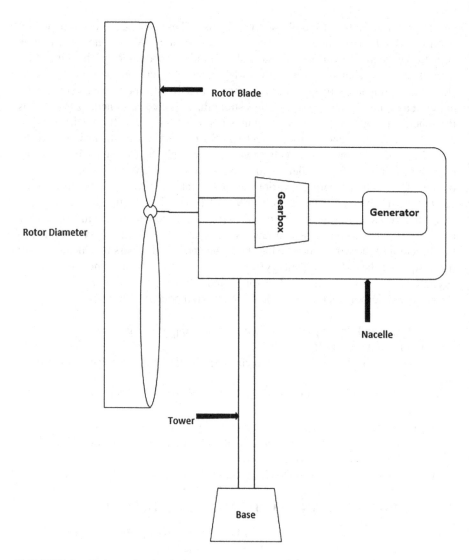

FIGURE 2.1   Horizontal-axis wind turbine.

2.5   WIND SPEED AND DIRECTION

For all well and it should be suitably possessing a shiny new WT on the outline of the floor or fixed onto the rooftop, but if the wind doesn't now blow for a long time, then there is no movement in the WT and it will not produce any power. Additionally, if the power in the wind is excessively high, then it will rotate the WT at very high speeds, which would destroy the WT or would produce overheat. Due to this, it will generate a high amount of output voltage or current. However, it is extensive at the same time while positioning a WT as part of a WE gadget to understand earlier how an extremely good wind would be available there and at how much speed the

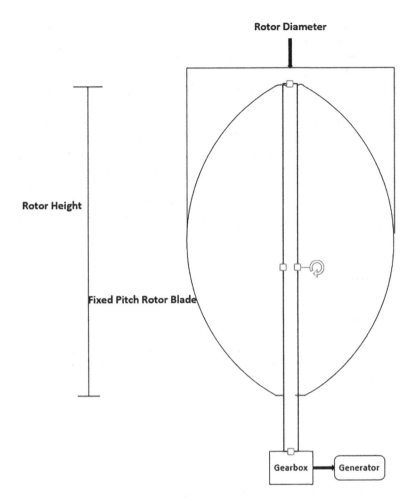

FIGURE 2.2   Vertical-axis wind turbine.

wind is actually blowing there [6]. The wind pace may be varying from very low to extremely high, i.e., due to storms at a particular site, so the speed of the wind is critical because of the reality that the amount of energy a particular windmill can produce is dependent on the actual pace of the wind. Wind velocity can be evaluated with the use of a weather vane or using a gauge called an anemometer. An anemometer is a device used to measure the rate and the amount of WE (wind energy) at a particular area. It consists of an arrow-long established metal or wood vane installed of a shaft immoderate inside the facet of the air and is outlined to issue inside facet in the direction of the wind.

The side from which the wind is blowing is called the direction of the wind. It normally has three arms with long cup established bowls at the peak location that spins nearby the shaft. These cups grasp the wind and passes it to the shaft to measure the speed of the wind. The shaft rotates faster, the harder the wind blows. This type of anemometer is used by maximum organizations as these are less expensive than others. A virtual circuit inside the anemometer computes the extensive types of cycles in accordance with minute and transforms that figure into miles (mph). Different formations of wind velocity are shown on the screen of the anemometer, or a hyperlink on the laptop proposes the physical motion of the wind. Any person can find out the charge and the amount of wind strength available very efficaciously with the use of a low-cost anemometer and a modern computer. All the available commercialized WMs have a minimum revolving pace at which they can produce electricity, which is known as the cut-in wind speed. Naturally, if the velocity of the wind is extremely low or zero, the potential of the WT for producing any beneficial output should be zero (0). In addition to this, even as immoderate winds revel in sturdy and incorporate big portions of available wind strength, the era of power that can be captured may be very small as the ones sturdy winds do now not stand up very often (which incorporates in storms).

2.5.1 POWER IN THE WIND

The power in the wind is measured in watts. It is also the power that can be withdrawn from the wind by our WT. Wind power withdrawn is determined from the size of the blades of the rotor, the speed of the wind, and the density of the air. Furthermore, theoretically the power required in transferring the air go along with the drift charge of kinetic electricity in keeping with the aid of using a wind turbine and is given with the aid of using the equation:

$$P = 0.5 \times \rho \times A \times V^3 \times C_p \text{ watts}$$

Here,
   P is called the wind power,
   ρ is called the density of the air in kg/m³,
   A is called the circular area swept by the rotor in m²,
   V is called the velocity of the air in m/s or mph, and
   C_p is called the coefficient of power (efficiency). It is the ratio of the power in the wind to the produced mechanical power, which should usually be around 35%–45%.

Looking at the equation, we see that the area of the rotor, i.e., in m², remains constant always and the density of the air depends upon the location and it should also be always constant. The power contained within the wind depends on the velocity of the wind. By simplifying the above equation, we are able to provide K. V³ in which K is hard and fast regular representing the mixed constant location of the blade of the rotor, mass of the air, and performance of the WT. Then, by this way, it means that the available strength of the wind is directly proportional to the cube of the velocity of the wind and this is very important as an insignificantly very small change in the velocity of the wind leads to a very high difference in the energy.

2.6   ADVANTAGES AND DISADVANTAGES OF WIND ENERGY

2.6.1   ADVANTAGES OF WIND ENERGY

- Wind power generation is a dirt-free and non-conventional energy genera-tion that doesn't liberate or emit pollutants into the ecosystem at any stage in operation as there are no chemical procedures involved.
- Modern mills produce little or no mechanical noise while running, besides a low "humming" sound.
- Wind power, that's truly a secondary element of sun's energy, is a "renewable source of energy" because there'll constantly be wind as long as the solar radiations in the daylight continue to warm the atmosphere of earth irregu-larly and the rotation of the earth continues as it is.
- Even though the strength in the wind fluctuates day to day, the whole out-put of strength over a set period of time fluctuates throughout most effec-tively as a small percentage, as these WMs are constructed and installed to respond at a speed of wind that lies normally in between around 12 and 70 m/h or 6–30 m/s.
- However, these WMs and largely the WFs (wind farms) need a large area of land for installing them; at the same time, the land under the WF can be used for the generation of wind power and some other uses such as planta-tion and animal grazing under the propellers of the WM.
- The production of wind power may be done in far-flung areas, and it can be from very small personal use and domestic use to absolutely large WF, so these are at large distances from the main road, i.e., on the mountains.

2.6.2   DISADVANTAGES OF WIND ENERGY

- There are many environmental impacts of wind power as many people reflect about consideration on WF, and these are not much attractive. The WM [7] may additionally have a horrific impact that has to be taken into consideration as a form of pollution.
- WFs additionally need a huge area of land or have to be located in environ-mentally sensitive areas, which include the wastelands, areas on top of any hill, the peak of any mountain, or offshore areas, i.e., near the sea, where the pressure and speed of the wind are very much influential and compatible.
- The WT has a huge propeller, and this calls for the KE (kinetic energy) of the wind to turn it. This means when there is no wind or a very small wind speed, then this WT doesn't rotate and also does not produce any electrical power.
- The WFs also have impacts on flying birds as these sometimes injure or kill the birds and sometimes also change the flying style of the travelling birds and hunting birds. There are a small number of other birds that are flying in the night, e.g., bats; they are also injured and killed by the blades of the WT.

- WMs also create noise pollution even though they produce very small humming noise. When these blades rotate, there is often a sound of very small frequency because these blades of the WT rotate at a fast rate.
- The initial cost of these WM plants is also very high because of the cost of transportation and other works, land cost, etc. All these things make wind power generation very costly than other power generation methods using fossil fuel.
- The plants for changing wind power into electricity are installed far away from populated city regions, which means that the power has to be transferred from these locations to the grid by means of very lengthy cables.
- Even though we cannot predict the annual variation in wind, we can only make an estimate of it and it is completely unpredictable. The wind in any area changes every hour, and also the generation of electricity in the area is also not the same.

2.7 WIND SPEED

Wind power has become popular over the previous couple of years or so, and we can see wind generators organized collectively on land and offshore, which can be acknowledged as "wind farms," to seize massive quantity of energy straight away and feed it into the country's grid. But the question is whether there is sufficient wind velocity to run those massive windmills to generate energy. Previously, the wind's power was used to drive cruising boats or operate windmills for irrigation and milling; however, at present, with improvements in cutting-edge turbine and generator technology, we are able to use the wind's power to energize our home and even cities. Wind power systems are one of the cost-effective energy generation methods; the renewable energy system converts the kinetic energy of the wind into mechanical or electrical energy that may be harnessed for realistic use. Depending on your wind resource, a small windmill should appreciably decrease your electricity bill. But the wind blow must be tough and continuously sufficient at your location to make the funding invested in a small wind turbine machine economically viable. There are many distinct kinds of wind turbine designs that can be of use in domestic, camping, and farmstead applications, which can be both on-grid and off-grid. However, they are not suitable for every place because the wind speed varies every day. People who stay close to the oceans are well informed of the pressure of the wind velocity and the way it influences their lives; while the ones staying in cities are informed of the velocity of the wind only when there are rains and storms. Wind velocity is pretty certainly the velocity at which the wind moves through the ecosystem and as such may be measured. When we're searching for a brand new residence, we're continuously being told the location of that place, which is also the same for wind power. The place of the wind turbine is critical, as there are many regions around each country where the wind blows surely non-stop, 24 hours, 7 days a week. At the same time, there are also different regions that get hold of very little wind so that it will generate enough quantities of power and be a cost-efficient renewable energy resource. A wind turbine generator desires to be at the proper place. The wind power can vary extensively over a small region; an area only some hundred yards away can see large variations in wind flow, which is recognized as "wind speed." This difference is specifically due to the form of the nearby terrain, hills, mountains, and probable obstacles such as trees and homes, which might be present in the wind's path. The wind possesses power through distinctive features of its movement, and this form of electricity—energy of the movement—is known as kinetic energy in which turbines are used to supply power. The total energy output obtained in a day is immediately associated with the wind speed at that definite location. Wind pace in any location is tremendously variable, and despite the fact that we cannot see the air shifting, we will nonetheless sense it. So any wind power tool able to slow down this mass of shifting air can extract some part of its kinetic energy and convert it into beneficial energy.

2.7.1 Wind Water Pumping System

Water pumping by wind with the use of windmills and generators is certainly considered one of the earliest inventions of humans. Historically, the electricity produced from the wind has been used in a huge variety of wind-powered applications, starting from grinding grains to sawing wood. There are many hundreds of thousands of humans in the world who do not have access to freshwater for their daily needs. In a lot of situations, water has to be obtained from wells or aquifers. But to make this water accessible, it first needs to be pumped to the floor from these sources. Stand-alone wind power plants are a suitable strategy to provide smooth energy to off-grid clients in remote places, letting them be absolutely unbiased from any oil price fluctuations. A normal, stand-alone wind power plant commonly consists of the following:

- One or more than one wind generator to harness the wind's power. These can vary from some watts (for micro-, mini-, and small structures) to numerous kilowatts based on the demand of energy and the accessible wind potential.
- A suitable electricity storage device. Usually, this can be a lead–acid battery or an array of batteries to permit a range of hours of self-reliance for the time when the wind does not blow.

In addition providing sufficient quantity of power to charge battery banks, wind generators can also be used to pump water. The majority of wind generators constructed in the starting stages were used for non-electric works. Wind water pumps are a mechanical appliance situated at some height on top of the pinnacle of a wooden tower and it will pump and discharge the water in the land to make it wet and useful for agriculture. After all, wind generators cannot absorb any water and this makes them best to be used in dry or drought-afflicted areas. Nowadays, automatically pushed water pumps are nonetheless an amazing feasible option; however, with advances in generation, there are also some large amounts of different feasible wind power approaches that require shaft strength also. They must include wind electricity-powered pumps for water [8] and a traditional water pump run by wind mounted up in a mixed PS (power system). The most commonly used type of wind-driven method for water pumping is certainly of mechanical type. Typically, the well-known water pumping technique using WE must include the following things: the rotor for the wind, tower for height, a pump, i.e., mechanical pump, a linkage that is also of mechanical type, a stream which should be absolutely furnished with water (or some other sources of water, such as a pond), and at last, a water pipe for delivering the pumped water. Otherwise, there can be a water storage facility adjacent to the water pump. Most of the water pumps in operation that use WE at present are of diverse structures such as piston pumps of reciprocating type and rotational pumps of screw kind type. In the piston pump of reciprocating type, the WT is attached to a crankshaft and gearbox, which transforms its rotating motion into up–down motion and is attached to the piston over the pump on the lower part of the pipe of a pond or well. Through the following half of the cycle, the piston starts moving down, which causes the valve of the piston to open and the water starts to flow through the top of the piston that is once more geared up again within the next half of the cycle. The amount of water superseded in the course of every knock is based on the diameter of the piston, which is similar to the inner diameter of the cylinder, and the stroke duration. Such an inconstant motion of the pump and piston suggests that the water is not flowing stably, but it can essentially be a pulsatile motion due to the reciprocating action of the water.

2.8   WIND ENERGY GENERATORS

The most important component of WE generation is the electric generator. The blades of the WM transfer the KE of the wind to rotating energy, and the purpose of WT generators is to convert the mechanical energy into electrical energy. After that, it is supplied to the grid. The frequency of the generated electric current is also matched with the frequency of the grid. A WT is composed of essential components. A low-speed electricity-powered generator is utilized for converting the mechanical power generated by the WE into electrical energy or power for energizing our homes and for doing some other useful works in industries.

1. Direct current (DC) generator, also called a dynamo.
2. Alternating current (AC) synchronous generator, also called an AC generator.
3. Alternating current (AC) induction generator, also called an alternator.

2.8.1   DC GENERATOR

DC generators can be used for the generation of electricity from the wind. A DC generator used in wind PPs has the following components: a WT, a DC generator, which can be either self-excited or separately excited, an IGBT (insulated-gate bipolar transistor) inverter, a step-up transformer, a controller, and an electrical grid [9] for supplying the generated power to the grid. In DC shunt-wound machines, as the operating speed of the DC machine increases, the current in the field windings of DC machine will grow and at the same time the torque that drives the WT determines the actual velocity of the WT. The commutator used in the DC machines converts the generated AC power into DC power, and it should be collected at the carbon brushes as an output. Due to the use of commutator and carbon brushes at the output, they need regular maintenance, and because of this reason, these are very costly for WE generation.

2.8.2 AC Synchronous Generator

Other types of generators used in WTs are AC synchronous generators, but these require power supply for their excitation; that is, they should be either DC excited or AC excited. They use either permanent magnets or electromagnets. Because of this reason, these are called the PMSG (permanent magnet synchronous generator) and EESG (electrically excited synchronous generator), respectively. The electrical power is produced in the three-phase stator winding of a generator when the rotor is driven by the WT, and this three-phase power is connected to the grid by the use of a power converter and a transformer. In case of constant-speed SG (synchronous generator), the speed of the rotor should always be equal to the synchronous speed; if this does not happen, then the machine will run out of synchronism; i.e., the synchronism will be lost. Tower shading effects occur in the constant-velocity SG, and they may produce random wind fluctuations and disturbances. Furthermore, when the WT generators will be predisposed, there will be a very less damping effect because these don't allow electrical transients to be absorbed. The frequency of the grid should also be matched with the frequency of the generation of the SG, which calls for an elegant operation. The disadvantage of these SGs is that they are very composite and very much costly, and the susceptibility of failure is very high as compared to IGs (induction generators). Due to the very low mass and excessive electrical density, these PMSGs are used by many organizations for more than decades. The shape of permanent magnet turbines is extraordinarily straightforward. To generate an ordinary magnetic location, the uneven PMs (permanent magnets) are installed on the rotor, and the power is accrued from the stator by using SRs (slip rings) or commutators or by using brushes. These PMs are occasionally integrated right into a tubular type of stable rotor, which is often made of aluminum to reduce the overall cost of the generator. The essential utilization of PMSG is the same as that of the SG because we can drive the PMSG asynchronously. Because of the same benefits such as the exclusion of the commutator, SR, and brushes, these PMSGs are robust, dependable, and very easy to understand. One disadvantage of this PMSG is that it cannot generate electricity at a fixed frequency when there is variation in the actual wind velocities. To avoid this disadvantage, the WT should be connected with the electrical grid with the help of rectifiers, i.e., AC–DC–AC [10], and converters. That means AC electricity is firstly rectified into a fixed DC and then it is again converted into AC. In addition to this, these PMSGs are capable of directly driving appliances, casting off the inconvenience of using gearboxes because they contribute to majority of failures in the WM. Only one prospective version of SG is the immoderate temperature generator with high superconductivity. These types of generators with very high superconductivity have modules with copper windings on the stator; coils on the field winding; support structure for the rotor; cooling system for the rotor, core of the rotor, and back iron on stator; etc. These coils made of very high superconducting material are capable of delivering current up to ten times greater than the normal or simple wires of copper, with slight resistance and losses in the conductor.

2.8.3 INDUCTION GENERATORS

IMs (induction machines) are sometimes used as IGs (induction generators). So, an IM behaves as an IG when:

    i. Slip becomes negative; due to this, the rotor current and rotor EMF attain negative values.
    ii. The prime mover torque becomes opposite to electric torque.

Suppose an IM is coupled with the prime mover whose speed can be controlled. If the speed of the prime mover is increased such that the slip becomes negative, the speed of the prime mover becomes greater than the synchronous speed. Due to this, all the conditions mentioned above will become fulfilled and the machine will behave like an IG. Now, if the speed of the prime mover is further increased such that it exceeds the negative maximum value of the torque produced, then the generating efficiency of the generator vanishes. Clearly, the speed of the IG during the whole operation is not synchronous; therefore, the IG is also called an ASG.

An IG is not a self-excited machine; therefore, in order to develop the rotating magnetic field, it requires magnetizing current and reactive power (Q). The IG obtains its magnetizing current and Q from various sources such as the supply mains or another SG. An IG can't work in isolation because it continuously requires Q power from the supply system. However, we can have a self-excited or isolated IG if we use a capacitor bank for Q power supply instead of an AC supply system. During generator operation, the prime mover (turbine or engine) drives the rotor past the synchronous speed. The stator flux still induces current in the rotor, but since the opposing rotor flux is now cutting the stator coils, an active current is produced in stator coils and the motor now operates as a generator, sending power back to the electrical grid.

Now, the next question is how it works. Consider an AC supply that is connected to the stator terminals of an IM. The rotating magnetic field produced in the stator pulls the rotor to run behind it, and then the machine acts as a motor. Now, if the rotor is accelerated to the synchronous speed by means of external prime mover, the slip will be zero and hence the net torque will be zero. The rotor current will also become zero when the rotor is running at the synchronous speed. If the rotor is made to run at a speed higher than the synchronous speed, then the slip becomes negative. A rotor current is generated in the opposite direction because the rotor conductors cut the stator magnetic field. This generated rotor current produces a rotating magnetic field in the rotor, which pushes (forces in opposite way) the stator field. This causes a stator voltage that pushes the current flowing out of the stator winding against the applied voltage. Thus, the machine is now working as an IG. The IGs are of two types:

    1. Self-excited induction generators (SEIGs) with squirrel cage rotors and
    2. Doubly-fed induction generators (DFIGs) with wound rotors.

An IM needs Q power for excitation, regardless of whether it is operating as a motor or a generator. When an IG is connected to the grid, it takes Q power from the grid.

But, if we want to use an IG to supply a load without using an external source, e.g., grid, a capacitor bank can be connected across the stator terminals to supply Q power to the machine as well as to the load. When the rotor is rotated at enough speed, a small voltage is generated across the stator terminals due to residual magnetism. Due to this small generated voltage, a capacitor current is produced, which provides further Q power for magnetization. IGs are suitable for wind-generating stations as in this case the speed is always a variable factor. Unlike SM, IGs are load dependent and cannot be used alone in grid frequency control. SEIGs are the generators which are not connected to the grid. These are stand-alone systems. In the future, there is going to be lot of distributed or decentralized power generation. This is because when the places are remote, then we would like to have a decentralized generation or distributed generation. If we connect an appropriate three-phase capacitor bank across an externally driven IM, an EMF is induced in the windings due to the excitation provided by the capacitor. This phenomenon is termed as capacitor excitation. The essential conditions for self-excitation are that there should be remnant flux in the machine and the charge in the capacitor should initiate the excitation. On the other hand, the IM can be doubly fed if we make the rotor terminals accessible and the rotor terminals can be made accessible only in the wound rotor IM where the rotor winding is identical to the stator winding. The stator winding is made for three phases and P poles; similarly, the rotor winding is also made for three phases and P poles. It cannot be wound for any other poles. A DFIG provides better control on speed as compared to SEIG.

2.9 CONCLUSIONS

So, with the advancement in technology, the increased demand for electrical energy and depleting fossil fuels have caused us to move toward renewable energy resources. Wind energy is a freely available energy with which we can generate power easily and feasibly. By using induction generators, the efficiency of the system can further be increased. Lots of research work can be done in this area.

REFERENCES

1. International Energy Agency (IEA), Energy Technology Perspective 2016.
2. R. Thresher, M. Robinson, and P. Veers, (2008), Wind energy technology: Current status and R&D future, in: *Physics of Sustainable Energy Conference*, University of California at Berkeley, pp. 1–24.
3. I. Pioro and R. Duffey, (2015), Nuclear power as a basis for future electricity generation, *Journal of Nuclear Engineering and Radiation Science*, 1, 1–20.
4. A. Kumar, M. Z. Ullah Khan, and B. Pandey, (2018), Wind energy: A review paper, *Gyancity Journal of Engineering and Technology*, 4, 29–37.
5. A. Khandakar and S. B. A. Kashem, (2020), Feasibility study of horizontal-axis wind turbine, *International Journal of Technology*, 1, 140–164.
6. D. Sangroya and J. K. Nayak, (2015), Development of wind energy in India, *International Journal of Renewable Energy Research*, 5, 1–13.
7. M. O. Apunda and B. O. Nyangoye, (2017), Challenges and opportunities of wind energy technology, *International Journal of Development Research*, 9, 14174–14177.

8. D. Srikanth, K. Himaja, Ch. Swasthik, and P. Uday Kumar, (2020), Water pumping system using solar and wind power, *International Journal of Engineering Research & Technology (IJERT)*, 613–615.

9. H. Ma, L. Chen, P. Ju, H. Liu, N. Jiang, and C. Wang, (2009), Feasibility research on DC generator based wind power generation system, in: *1st International Conference on Sustainable Power Generation and Supply, SUPERGEN'09*, pp. 1–5.

10. D. Prashar and K. Arora, (2020), Design of two area load frequency control power system under unilateral contract with the help of conventional controller, *IJICTDC*, 5, 22–27.

3 Flexible Load and Renewable Energy Integration with Impact on Voltage Profile of a Large Size Grid

Bhavesh Vyas and Vineet Dahiya
K. R. Mangalam University

M. P. Sharma
Rajasthan Rajya Vidyut Prasaran Nigam Limited (RVPN)

CONTENTS

3.1 INTRODUCTION TO EXISTING SCENARIO

Due to the incorporation of high renewable energy resources, the prerequisite of reactive power reimbursement of the grid changes, as per load variations consuming the power. It indirectly affects the voltage and power factor for grid operations. It results in frequent breaking down/shutdown of conventional power plants, addition of distributed generation in power system [1–5]. Thus, disparity in grid operations strategy has led to multiple system effects and deterioration of power system. To maintain interconnected high-voltage network with the grids of other states, they have to be dynamically planned, or as confined by the utilities. Therefore, the work plan as per the literature available due to the latest research or that based on IEGC,

DOI: 10.1201/9781003242277-3

CEA, or any other central-level organization guidelines need to be followed for regular power system operations. The One Sun One World One Grid vision has raised renewable energy generation plans from the Indian states having high potential for renewable generation through a common new transmission corridor. The nation's work for achieving a target of 175 GW by 2022 is sped up. But along with this, studies must be carried out in parallel to extrapolate the possible pros and cons of renewable energy addition to the network. The work discussed hereby shows a study conducted at the State Electricity Board Rajasthan (SEBR) as a test network to identify the effects of renewable energy addition with dynamic behavior of load.

Since the national grid is in synchronous mode with links of inter- and intra-tie line ends, maintaining the system inertia or voltage as per IEGC or CEA rule is well explained for complete EHV transmission corridor, but still discrepancies prevail. It increases the need for planning at GSS level. Therefore, mathematical analyses to identify system deficits are carried out initially in this chapter, and then multiple power flow studies are done to identify the test network requirements. Feeder connectivity relations-based inductive VAR lacks are systematically weighed, and predicted for SEBR by seeing possible fully compensated VAR structure [6–11].

The renewable energy penetration in SEBR is not dominating presently; it accounts for nearly 30%–33%, but as per planned, allotted schemes, a contribution of 66%–70% is expected in the upcoming few years. So, operation schemes designed at present will be the ruling the one in future after the commissioning of renewable energy setups. The addition and removal of VAR compensation devices needs to be planned for the structure at load dispatch levels to grid substations. Artificial intelligence and machine learning could take only a part unless and until the modernized sensor-based grid is available, which may take long time to achieve considering the present-day setup taken for observance. The process explained hereby signifies the participation of network elements in maintaining proper system flows at the required time under National and International defined limits.

The following objectives were decided to be fulfilled from the load flow studies carried out.

- To identify and classify uncompensated grid substations of 400 kV level, which are violating the IEGC/CEA-defined planning standards of voltages.
- To obtain shortages in inductive VAR compensation at 400 kV GSS by the proposed mathematical analysis.
- To prepare a model of SEBR with varied load and varied renewable energy penetration to identify system supplies.
- To demonstrate variation in voltage profile with increasing renewable penetration from 20% to 70%.

Conclusive assertions based on comparison of swing generator flows, and power losses as per the situation framed is provided.

Based on the findings, the enumerated steps could be suggested to the utilities for considering grid operations on daily basis. Optimum utilization at the time of availability is the basic constraint of renewable energy.

Also, if future point of view with respect to electric vehicles is discussed, each single unit of EV must be in the range of 10 to 8 kW. Accordingly, the load growth will definitely be high as per the statistics more than 10 million EVs will be added to contribute future plan of zero carbon emission environment.

Hence, the growth in load will be in multiple formats, and during those times, the renewable energy could be channelized. Thus, power system planning plays an important role and it should be carried out with simulation practices, extrapolation based on mathematical techniques, etc., for every 5–10 years. The next section will thoroughly detail the existing problems in the grid.

3.2 PROBLEM IDENTIFICATION

The SEBR regularly records the data of the existing financial year and others so that discussions and decisions could be made as per the requirements.

Table 3.1 shows that the SEBR receives a contribution of 11.02% from wind and 14.30% from solar, and overall, 10.81% as a whole. Thus, suggesting increment in renewable addition to conventional set up. Similarly, recorded meter readings from all databases are analyzed and a graphical chart for 400 kV voltage is plotted in Figure 3.1.

It could be inferred that majority of high voltage peaks are recorded throughout the year 2018–2019. Peak voltages exist at all GSS. High voltages cause pre-aging of devices, which in turn causes earlier repair and maintenance issues in network

TABLE 3.1

Statistics of Installed Capacity of Solar and Wind Generation till March 31, 2021

S. No.	Test Network (till March 31, 2021)	Wind Power	Solar Power	Total Capacity
		All in MW		
01	Rajasthan state	4,327	5,733	10,205
02	All states total	39,247	400,085	94,434

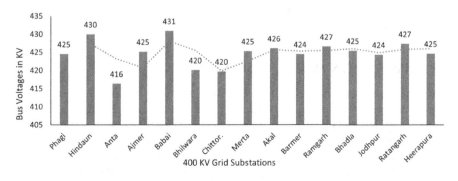

FIGURE 3.1 Month-wise average of bus voltage for 400 kV GSS of RSPS.

TABLE 3.2
Allowed Voltage Limits as Per IEGC and CEA Standards

Ideal Voltage Range	Peak Voltage Level Limit	Low Voltage Level Limit
400	420	360
220	245	200
132	145	120
66	72.6	59.4
33	34.98	30.03
11	11.66	10.01

elements. Oversaturation of the transformer's core starts rising along with an increase in high impact over the swing generator.

When compared to the above chart values, the SEBR has already finalized the voltage limits as per the instructions from standard bodies, which are shown in Table 3.2.

Based on the system position, multiple stages are assumed so that the state of more than 90% of buses could be identified as level 1, 2, or 3. The last one will be the critical stage after which load rejection-type regulations need to be executed for the safety of generators and the system.

As it could be understood, equalized three-phase voltages should be there; otherwise, any deviation in between them will reduce the system performance by introducing circulating type negative torque; overheating may occur, which could result in the malfunctioning of relays or equipment linked to the network.

The companies assigned for a particular work need to monitor grid parameters timely, and if not found satisfactory, then they must take actions or impose penalties for the sake of improvement. Indian standards took reference of voltage regulations from IEEE 519-2014, EN 50160, and IEEE 519-1992 to lay down voltage levels. In many states, at present either no limits are finalized, or if finalized, then no monitoring is there to synchronize the system with National Synchronous Grid operating. Even some states have restored values of voltage limits different from central commands; hence, these issues will become vulnerable in smooth grid operations sooner than later. All these deficiencies cause abrupt tripping, outages, rise in system fault levels, etc. Thus, a strategy to maintain highly recorded voltages as per IEGC standard should be formulated for future purposes.

3.3 SIMULATION MODEL DESIGN

The State Electricity Board Rajasthan (SEBR) is the test network considered in this chapter. It is positioned in northern part of India with one international boundary at the west and the rest being national boundaries from five different states. At present, it is the biggest state of India and the seventh largest populated area (approximately 8.1 crores till May 31, 2020, as per UID records). It covers an area of 342,239 km², which is 10.4% of the total area of the country. The SEBR has a huge network voltage hierarchy from 765 to 33 kV in transmission and distribution. If source-wise generation is discussed, then the contribution of coal-based thermal power plant is 57%, where the generation of voltage varies from 11 to 33 kV as per generator unit size and unit running time period is referred. The particulars of the Rajasthan State Power System (RSPS) grid substations involved in energy activities are briefed in Table 3.3. A total of 606 number of grid substations are actively taking part in the power transmission and distribution activities. The actual source-wise contribution is provided in Table 3.4. RSPS is having more than 0.39588 lakh circuit KM area covered till March 31, 2019, database locked now. The complete network is modeled in a single system without isolating. All state entities work in cooperation with central and private authorities for the successful operation of the grid. Due to the state's large area, wide EHV transmission corridor exists here. So, present as well as future planning plays a great role in VAR management on need basis.

The contribution of renewable energy will increase steadily, and it has to be raised to 80%–90% range to achieve the lowest emission of carbon constituents into the environment. Source-wise generation with reference to Rajasthan state network is summarized in Table 3.4.

The energy map drawing detailing the network updated till locked date is given in the map shown Figure 3.2.

The addition and removal of elements to/from the grid is instructed by the load dispatch and monitoring control room [12–16]. The modeling of test network with

TABLE 3.3
State Electricity Board Briefings of Grid Substations

Particulars of GSS	No. of GSS/MVA Capacity till March 31, 2019			
	RVPN	PPP (RVPN)	PGCIL	Total
765 kV	2 (7,500 MVA)	–	2 (6,000 MVA)	4
400 kV	16 (13,125 MVA)	2 (1,260 MVA)	9 (7,855 MVA)	27
220 kV	121 (30,035 MVA)	1 (260 MVA)	–	122
132 kV	437 (31,420 MVA)	15 (425 MVA)	–	453
Total	**576**	**18**	**11**	**606**

TABLE 3.4
Source-Wise Power Generation Contribution

Type of Generation		Source-Wise MW Generation (MW)			Source-Wise %
Hydropower		–		1,961.954	9%
Nuclear		–		456.740	2%
Thermal	Coal	11,181.496	53%	12,006.096	57%
	Gas	824.600	04%		
RES	Wind	4,139.200	20%	6,652.850	32%
	Biomass	101.950	0.48%		
	Solar	2,411.700	11.44%		
Total				21,077.64	100%

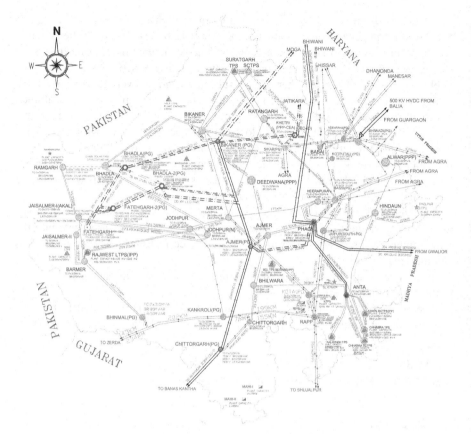

FIGURE 3.2   RSPS network drawings depicting 400 kV nodes and above.

12,874 MW as active load and 2,470.032 MVAR as reactive load is carried out in MiPower database modeling application. The network is composed of generator buses and load buses with a total count of 1,015 bus constituents (765 kV—15 Nos, 400 kV—59 Nos, 220 kV—207 Nos, and 132 kV—713 Nos). Along with 1,256 connecting lines, the substations are supported with 226 transformers as voltage stepping devices. By setting all the reduction factor parameters and defining the zones as such, SEBR is divided into three different DISCOMs. These zones are modeled according to the contribution provided by solar generation, wind generation, etc. Network elements such as shunt reactors (180 Nos) and shunt capacitors (449 Nos) are also installed in the network. After preparing a detailed database of the complete network, the final GUI-based simulation model is prepared for the test network. Different zones made are shown in Table 3.5.

Swing bus is operated as an imaginary generator at the location shown in the power map. Since majority of the network is in synchronism with Northern Regional Grid division, the swing bus point is modeled by keeping the same ideology. The state tie line points will also be there near the swing bus. Since in real the test network is having connectivity to national grid at multiple voltage level, such as 765,

TABLE 3.5
Load-Based Zone Modeling to Provide in MiPower Software

		Connected Load		Load in 1 A	
		Active	Reactive	Active	Reactive
Zone No.	Defined Zones as Per Test Network	MW	MVAR	MW	MVAR
1	Jaipur	7,724	2,640	3,658	792
2	Ajmer	4,901	2,051	2,321	615
Total		**34,438**	**8,042**	**12,875**	**2,470**

400, 220 and 132 kV all are represented in our model. If the tie line substation points import power into the network, then they will be treated as a generator source. If these generating stations are getting power from the network, they will be treated as load points during the simulation study [17–19]. Multiple zones are created to study the existing setup, and the loads updated in the base network are applied in base case simulation modeling. Simulation design of the test network is carried out, and analytics are accomplished using MiPower software as follows:

- A test network of RSPS is drawn as per energy drawings, restricted at 132 kV with low RE mode.
- A replica of network power flows matching the existing situation is analyzed.
- The software-based simulator consists of 1,015 buses with 113 generator nodes, 1,256 transmission nodes, 698 load points, 180 SRs, 449 SCs, and one static VAR compensator mounted at 400 kV Kankroli GSS node.
- The consumption is represented at the 132 kV bus of each 400 kV node. The aggregate load in the test network is 12,874 MW and 2,470 Mega Volt Ampere ratings with 0.982 power factor as the load power factor.

Upcoming sections will detail the load flow solutions, mathematical analysis, and renewable energy integration results to compare and discuss the planning to maintain voltage profile [20–24].

3.4 POWER FLOW STUDIES OF MULTIPLE SIMULATED CASES

Power flow studies are supported to find the condition corresponding to 2022–2023 as per the network modeled. The simulated results of the test network in Graphical User Interface (GUI) file is displayed in the next section. Case divisions selected are as follows:

1. LFS for 12,874 MW Load of Existing Test Network with 20% Renewable Integration
2. LFS for 10,300 MW Load of Existing Test Network with 70% Renewable Integration

3. LFS for 12,874 MW Load of Existing Test Network with 70% Renewable Integration

4. LFS for 15,449 MW Load of Existing Test Network with 70% Renewable Integration.

The pictorial representation provides the visual classification from the database. It includes all the network constituents. Those are actually taken as a part of the network for modeling. Basic assumptions and consideration made for load flow study of the test network are furnished in Table 3.6.

The results from Newton–Raphson studies of the existing RSPS network are briefed in Table 3.6. Multiple cases are simulated, such as wind integration of 20%, 70% with varied load of 0.80 times down from base, and 1.20 times higher from base. The data updated till March 30, 2019, in the test network are applied for studies.

Multiple load flow studies are carried out to access the existing situation of the network and to obtain possible suggestions required after the identification and fulfillment of deficit VAR in the network. The voltages could be controlled if proper VAR levels are maintained with the need of the time approach. So, keeping a suggestive baseline of voltage limits under operations from CEA and IEGC, the parameters are fixed and shortages at GSS in terms of lacking VAR support are identified. The grid network presented in graphical user interface is best suitable on zero scale size layout. Exhibit images are obtained on 300 dot per inches that is suitable for chart size print. So that minor lines identification will be more viable. Zooming the images to 200–300% does not blur the image. The study as per power flow points could be easily carried out [25–28]. The (+) indicates the addition (generation) of power, and (−) indicates the withdrawal of power. The load flow result summary as per case variations is given in Figures 3.3–3.6.

TABLE 3.6
Load Flow Results of Multiple Cases Considered

Parameters		RSPS NR-LFS			
Load MF		**1.00**	**1.00**	**0.80**	**1.20**
RE %		**20%**	**70%**	**70%**	**70%**
Total real power load in MW		12,874.69	12,874.69	10,300.2	15,449.20
Total reactive power load in MVAR		2,470.03	6,925.77	1,976.02	2,964.03
Generation PF		0.938	0.721	0.578	0.973
Renewable generation	Wind	1,260.25	4,410.88	4,410.88	4,410.88
	Solar	4,197.16	8,575.08	8,575.08	8,575.08
Total shunt reactor injection in MVAR (-)		25,992.26	23,342.50	22,747.9	24,755.0
Total real power loss in MW		270.86	1,207.92	1,563.5	671.40
Percentage real loss		2.010	5.750	7.44	3.48

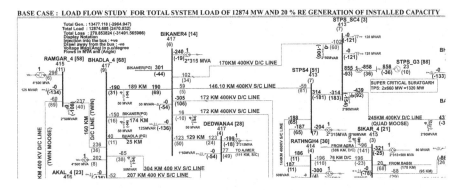

FIGURE 3.3 LFA of existing base case with 20% RE integration and 12,874 MW load.

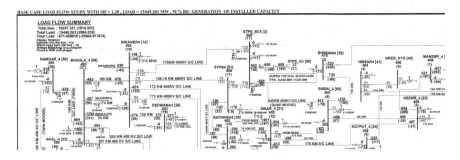

FIGURE 3.4 LFA with MF = 1.20, Load = 15,449.203 MW, and 70% RE generation of installed capacity.

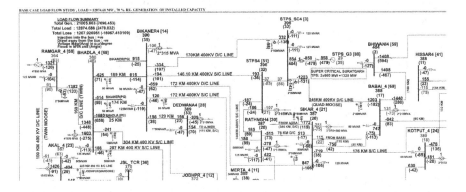

FIGURE 3.5 LFA with MF = 1.00, Load = 12,874.68 MW, and 70% RE generation of installed capacity.

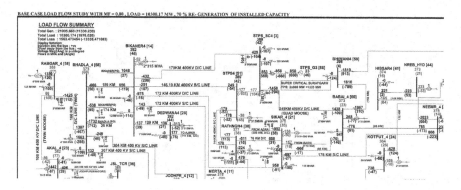

FIGURE 3.6 LFA with MF = 0.80, Load = 10,300.17 MW and 70% RE generation of installed capacity.

3.5 DYNAMIC SIMULATION STUDY COMPARISONS

Fluctuations captured by meters installed in the grid in terms of high voltage values with reference to dynamic load are utilized with the applied strategy. A variation of 0.80–1.20 times is taken from the base case to create peak and low load situation. Multiple cases are identified, and the voltage levels found are studied with a comparison chart including all the renewable integration cases.

As shown in Table 3.7, case variations are provided for each 400 kV GSS of RVPN concerning increased renewable penetration because VAR deficiency is obtained for RVPN network only. So, the cumulative increment in load is carried out to access the effects on voltage levels of grid buses.

The dynamic load of 2018–2019 base criteria (12,874 MW) is set based on peak recordings and electric power survey calculations. Multiple simulation cases are modeled with the analysis of various renewable energy additions into the test network. Substation-wise voltage profiles are graphed in the chart with variable renewable generation and load. So that the actual system demand could be identified.

Figure 3.7 shows the results obtained at Ajmer GSS (SS1): During 30% renewable integration, the voltage varies with low load toward high load stances. Although the bus voltage profile remains greater than 400 kV, this increases to 410 kV during 40% penetration and it shows the requirement of inductive VAR in the system. In case of 50%–70% renewable penetration, the voltage profile reduces from 410 to 388 kV in low load situations, but remains between 395 and 400 kV during high load conditions.

At Merta GSS (SS2), as shown in Figure 3.8, at 30%–40% renewable penetration, the voltage profile decreases from 408 to 402 kV. In case of 50%–60%, it is observed that as the load raises from case 5 to case 9, ups and downs are there, but at last, with on 70% insertion, voltages are reduced to very low levels (in case 5), which may lead the system toward low voltage trip situation, while it is not so at high load conditions.

As shown in Figure 3.9, at Heerapura GSS (SS3), during 30%–40% renewable integration, the voltage has to be managed with inductive VAR support, but higher loadings and higher penetrations are much more balanced. Low loading VAR management scenarios needs to be sought out by operators.

TABLE 3.7

Case Divisions as Per Variable Load and Renewable Energy Addition to SEBR

Case Division	Variations Undertaken	Active Load (MW)	Renewable Integration in %				
Case 5	0.80%	10,300.17	30	40	50	60	70
Case 4	0.85%	10,944.33	30	40	50	60	70
Case 3	0.90%	11,588.48	30	40	50	60	70
Case 2	0.95%	12,230.53	30	40	50	60	70
Case 1	**Base case**	**12,874.68**	30	40	50	60	70
Case 6	1.05%	13,518.84	30	40	50	60	70
Case 7	1.10%	14,163.00	30	40	50	60	70
Case 8	1.15%	14,807.16	30	40	50	60	70
Case 9	1.20%	15,449.20	30	40	50	60	70

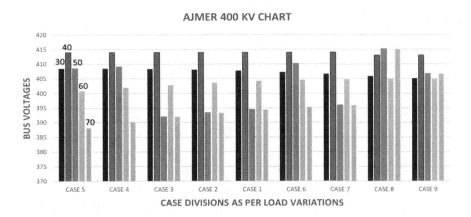

FIGURE 3.7 Chart 1 of SS1 indicating EHV profile.

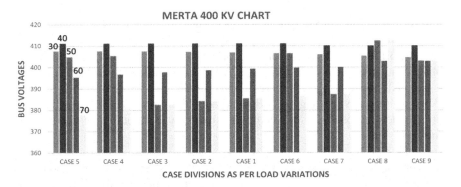

FIGURE 3.8 Chart 2 of SS2 indicating EHV profile.

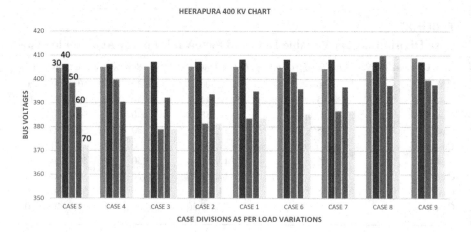

FIGURE 3.9   Chart 3 of SS3 indicating EHV profile.

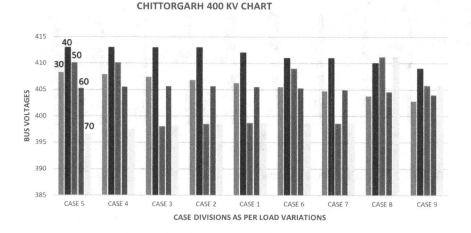

FIGURE 3.10   Chart 4 of SS4 indicating EHV profile.

Figure 3.10 shows the results obtained at Chittorgarh GSS (SS4): Chronicle high voltages violating planned IEGC limits exist when renewable penetration is 40%. Hence, VAR absorption is the desire for the overvoltage heading lines.

Bikaner GSS (SS5): It suggests that maximum high voltages exist during 50% and 70% of renewable penetration. Voltage will always try to shoot in high load and high renewable situation.

(Figure 3.11).

As elaborated in Figure 3.12, Bhilwara 400 kV GSS (SS6) submits a good range of voltage profiles with least deviation. The highest voltage recorded is 413 kV, while the lowest value obtained is 395 KV. Hence, the inductive VAR support is capable

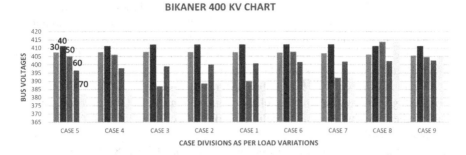

FIGURE 3.11   Chart 5 of SS5 indicating EHV profile.

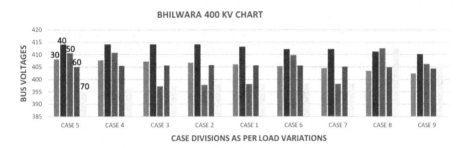

FIGURE 3.12   Chart 6 of SS6 indicating EHV profile.

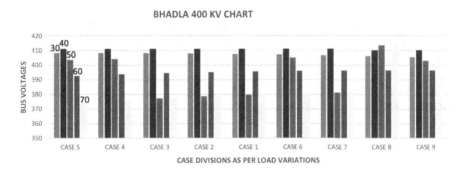

FIGURE 3.13   Chart 7 of SS3 indicating EHV profile.

of handling the inertia of maintaining balanced voltage profile, which is better as compared to other GSS.

The results at Bhadla 400 kV GSS (SS7), shown in Figure 3.13, suggest that as renewable integration is increased, the voltage level is reduced from 410 kV in case 5 to 370 kV. Renewable addition is supported till 30%–40% level, but when it is raised to higher voltage level, it gets disturbed due to variation in VAR levels of the network.

Figure 3.14 shows the results of Barmer GSS (SS8): It shows that the maximum voltage recorded is 420 kV, while the lowest is 371 kV; both are at low load situations. So renewable penetration during low load needs to be channelized or managed as per the need basis. Low-voltage facing substations must be prepared for recurrent problems, and as such, plans must be applied.

As detailed in Figure 3.15, at Babai GSS (SS9), it is suggested that up till 30%–40%, voltages are running in parallel under allowed range. But the situation changes when higher renewable energy penetration takes place. Lower voltages are recorded during low load situations.

Lastly, Figure 3.16 details the results of Anta 400 kV GSS (SS10). It suggests that with the variable renewable penetration, minimum voltage differences are recorded, but wide variations are there in cases when penetration is 40%, which is a case of low load situation. When the voltage variability from 411 to 402 kV is obtained, it is the highest of all scenarios observed.

As per the simulation study results variation in renewable energy integration along with load pattern is compared by chart in Figures 3.7–3.16, respectively.

Thus, from overall observations, it could be said that the utility has to add an inductive VAR support during low load conditions so as to manage the voltage profile within a desired limit. Also, the reactive VAR support must be maintained in

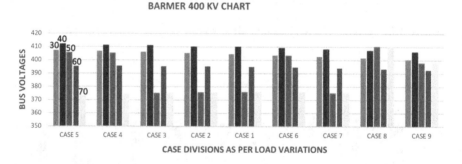

FIGURE 3.14   Chart 8 of SS8 indicating EHV profile.

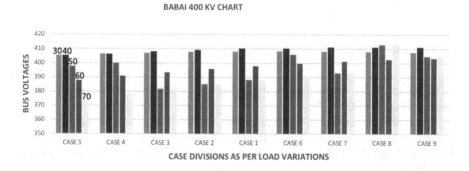

FIGURE 3.15   Chart 9 of SS9 indicating EHV profile.

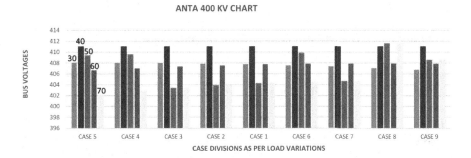

FIGURE 3.16   Chart 10 of SS10 indicating EHV profile.

such a way that high voltage peak surges do not occur in the system. These variable renewable penetration studies could be used as a reference by utilities for deciding action plans in similar situations that occur during EHV grid operations. Hence, higher load level needs capacitive VAR support, while low load conditions require inductive VAR support tools to maintain voltage levels. So, possible voltage control feature by inductive VAR support is required. It raises the requirements of advance algorithm-based network capable to predict dynamic variations and that can respond accordingly.

3.6   COMPARISON OF EXISTING GRID GENERATORS

An increase or decrease in load may affect the generator capability. NREB_TH is the swing bus. The higher the negative value, the larger the VAR drawl from the generators will be. It may cause the problem of thermal effects. The existing conditions of VAR drawl from all conventional grid generators are plotted along with load variations as 0.80, 0.85, 0.90, 0.95, 1.00, 1.05, 1.10, 1.15, and 1.20 times the base load, as shown in Table 3.8.

It could be discussed from the above table that if local VAR support is provided, then these large negative values could be reduced in a proper manner. Also, higher loading levels of generator could be reduced. Swing generation submits negative values in proportion to its increasing sequence. Similarly, after generators, the buses present in the network are compared by identifying their sequence in terms of the voltage profile they are attaining during normal grid operations. It is obtained that as the load reduces (overvoltage buses), the count rises to 609 numbers from a total of 1,015 buses. It suggests that at such a high count of case 5, these buses may lead to errors when reaching the 1.05 Per Unit range. So, care must be taken at grid level to maintain the specified voltage profile range of buses in a big network. Table 3.9 shows that during low load conditions, 60% buses are crossing limits or on a higher side, and in case of higher load conditions, 71 buses (approximately 7%) are crossing higher limits gained due to renewable integration.

Thus, to maintain voltage stability buses participation in stable range needs to be raised. VAR compensation should be done near them as per the needs.

TABLE 3.8
Generators VAR Drawl Comparison with Variable Load

Conv. Generator	C-5	C-4	C-3	C-2	C-1	C-6	C-7	C-8	C-9
BLTPS_G A	−29	−24.9	−20.2	−14.9	−9	−2.6	4.3	11.9	20
CHABR_G2 A	16.2	18.4	21.2	24.9	29.2	34.1	39.7	45.9	52.5
CHABRA_G A	−12.5	−9.8	−6.3	−1.9	3.3	9.2	16	23.5	31.5
KALINS_G A	−22.6	−17.2	−10.9	−3.7	4.5	13.3	23.1	33.7	44.8
KEWAI_G A	−20.3	−17.1	−13	−7.8	−1.5	5.6	13.7	22.7	32.2
KTPS#3 A	−22.6	−19.7	−16.3	−12.6	−8.5	−4.1	0.7	6	11.5
KTPS#4	−45.5	−39.7	−33.1	−25.7	−17.4	−8.6	1.1	11.6	22.7
RAPP#1	−24	−20.7	−17.1	−13.2	−8.8	−4.2	0.8	6.3	12
RAPP#2	−56	−48.9	−41	−32.4	−23	−12.8	−1.8	10.1	22.4
RAPP_D1	−37.9	−35.6	−32.4	−28.4	−23.6	−18.6	−12.8	−6.1	0.5
RAPP_D2	−37.9	−35.6	−32.4	−28.4	−23.6	−18.6	−12.8	−6.1	0.5
RAPPC_G	−17.9	−16.7	−15.2	−13.2	−10.8	−8.3	−5.4	−2	1.3
RJWESTG1	−29.9	−24.7	−19.1	−13	−6.5	0.5	7.9	16	24.2
RJWESTG2	−64.8	−56.9	−48.2	−38.6	−28.1	−17	−4.9	8.4	21.8
STPS_G	−49.5	−50.5	−48.7	−43.9	−36.2	−26.1	−13	3.1	22
STPS220G	−41.4	−36.5	−30.4	−23.2	−14.9	−5.6	5	16.7	29.6
STPS400G	−19	−19.3	−18.6	−16.8	−13.9	−10.1	−5.2	0.8	7.8
NREB_TH	−90.9	−226.5	−332.4	−408	−453.6	−469.8	−454.3	−407.1	−327.8

TABLE 3.9
Number of Buses Higher than 1.03 PU

Particulars	C-5	C-4	C-3	C-2	C-1	C-6	C-7	C-8	C-9
1.03 >	609	499	365	283	204	156	119	94	71

3.7   ASSESSMENT OF REQUIRED VAR SUPPORT

Volt-ampere reactive (VAR) power is a feature to maintain voltage levels in any grid. Presently ongoing grid operations of load sanctioning are not channelized as per dynamic variations. But the basic decisions for maintaining the voltage profile could be taken at utility end. The grid voltage is a cumulative result of coordination between generation and demand. Compensation should be both inductive and capacitive; that is, there must be proper availability of bus type and line type shunt reactors to control high voltages when unavailability of solar power or availability of excessive VAR is found in the system. Total compensation (100%) required by a line is identified from its length and per kilometer charging MVAr value. The shortage in VAR could be identified, and as per the need, equipment could be installed to fulfill shortages so that a balance of voltage profile working under planned limit could be obtained.

The absence of latest technological upgradations and lack of utilization of controlled algorithms to channelize the VAR flows have hindered the mounting of VAR compensation units. A general procedure on requirement basis is currently being followed, and that even comes under action after several power disruptions. Hence, to point out the level of VAR requirements, the degree of penetration at system level must be found out. Thus, the required VAR between transmission ends is obtained through following the steps provided in the flowchart.

Calculations are applied to all the GSS of state electricity board (RVPN). VAR deficit is obtained for 400 kV transmission line capacity. By differencing reactive power compensation degree of existing line with 100% compensation, the system compensation requirements could be identified. The flowchart presented in Figure 3.17 provides the steps to identify the existing demand of inductive VAR support required to manage the voltage profile.

After observing the "over- and under-compensation levels" of a grid. The strategy to maintain regular operations must be envisaged. The larger the distance from generation ends, the higher the VAR support desired, so the only way to handle voltage variations and fault occurrence is to support the grid with dynamic VAR devices. Power demand of Test Network substations are shown in Table 3.10. After fulfilling deficit VAR, the utilities are capable to manage voltage profile whose results will be transferred to lower voltage levels of the network.

Fulfilling this obtained demand shown in Table 3.10 Identifying the procured load as shown in Figure 3.10, it may maintain the voltage profile to level over a duration of time but this is not the ultimate solution grid operators have to be upgraded to an extent where necessity could be managed on daily day ahead basis based approach

3.8 CONCLUSIONS

From simulation studies, the existing voltage profile is obtained and cases of violation of the voltage profile at respective GSS are obtained. Deficiency in terms of inductive VAR support is obtained at 400 kV GSS transmission level. Simulations of multiple cases having dynamic variations (load and renewable penetration) are created from case 1 to case 9; that is, the load is from 10,300.17 to 15,449.2 MW from a base load of 12,874.68 MW. Along with this, renewable penetration from 30% to 70% has been simulated and studied. The base load has 20% renewable integration. From the studies, the following conclusions can be drawn:

- Out of the 16 400 kV grid substations, 12 GSS provide high voltage conditions. As a result, a violation of the IEGC and CEA rules is seen.
- Shortage of inductive VAR support is assessed for all substations operating at 400 kV voltage level.
- If steps are taken to provide 100% VAR support from GSS ends, then voltage control from utility end would be more reliable and balanced.
- The ratio of buses participating in high voltage region is higher during low load situation; also, generator VAR drawl contribution increases in case of dynamic variations.

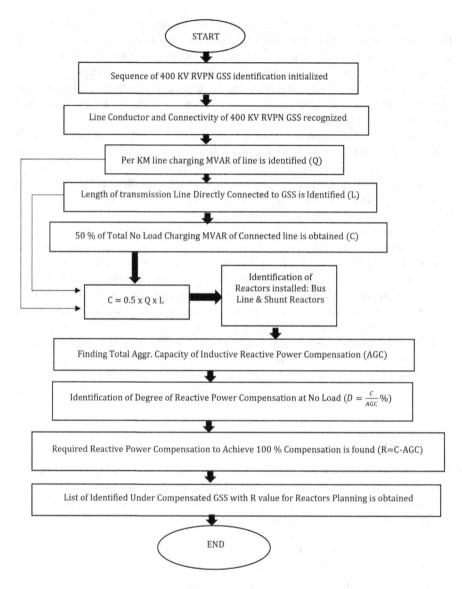

FIGURE 3.17 Flowchart to identify system VAR requirements to maintain voltage profile.

- Abruptly switching off capacitive VAR banks during night hours without adding the same amount of inductive VAR support may cause high voltage issues at present, and this may occur in future, too, unless proper actions are taken.
- Dynamically predicted scenarios as faced by utilities are picturized. By trying to suggest the possibilities of improvement to maintain the caused situation. Aroused situation deals with the higher renewable position in the grid, that will be a futuristic scenario.

TABLE 3.10
Demand in VAR Return at Identified Substations

Sub-station Details	Compensation Deficit	VAR Deficit VAR Req.
ANTA_4	211.22	141
HEERAPURA	116.26	114
MERTA	143.93	148
BIKANER	186.30	180
BHADLA	77.57	83
BARMER	70.57	71
AJMER	236.12	236
BHILWARA	110.37	48
BABAI	80.07	80
CHHITOR.	53.76	46
JODHPUR_N	130.33	128
JAISALMER-2	40.71	41

ACKNOWLEDGMENTS

The authors would like to thank the state electricity board RVPN and their university staff for providing the required support. Also, they acknowledge the software support provided by PRDC Pvt. Ltd. This is a non-sponsored work carried out to provide utilities a reference from planning point of view.

REFERENCES

1. Press Information Bureau, Government of India, Green Energy Corridor Report, Last Accessed on: 20.02.2021, https://pib.gov.in/newsite/PrintRelease.aspx?relid=116890, [Green Energy Corridor Project].
2. Press Information Bureau, Government of India, Ministry of New and Renewable Energy, India's First ISTS connected Wind Power Project Commissioned- Government's plan of 175 GW RE by the year 2022, https://pib.gov.in/newsite/PrintRelease. aspx?relid=183044, [ISTS].
3. S. Bajaj, Smart Grid Optimization supporting, National Electricity Plan, Clean Energy news and insights, https://mercomindia.com/india-39-billion-transmission-infrastructure-revamp/, March 18, 2019.
4. Central Electricity Authority, Government of India, Reports and Regulations, Transmission planning criteria. Manual last updated on January 13, 2019, http://cea.nic.in/reports/others/ps/pspa2/tr_plg_criteria_manual_jan13.pdf.
5. Central Electricity Regulatory Commission, Government of India, Ministry of Power, Indian Electricity Grid Code, old and revisions. Available online at, http://powermin.nic.in/sites/default/files/uploads/Indian_Electricity_Grid_Code.pdf.
6. State Electricity Board Rajasthan installed capacity of Grid Interactive Renewable Power as on 31.03.2021, https://mnre.gov.in/the-ministry/physical-progress.
7. Rajasthan Electricity Regulatory Commission (Transmission Licensee's Standards of Performance) Regulations, 2004, Published vide Notification No. Secy/RERC/REG/No. 24, dated 20.4.2004, http://www.bareactslive.com/Raj/rj1253.html.

8. P. Kundur, N. J. Balu, and M. G. Lauby, *System Stability and Control*, EPRI power system engineering series, Tata McGraw Hill publications, New Delhi, 2007.

9. R. N. Nayak, Y. K. Sehgal, and S. Sen, EHV transmission line capacity enhancement through increase in surge impedance loading level, 4 pp, 2006. doi:10.1109/POWERI.2006.1632620.

10. L. R. de Araujo, D. R. R. Penido, S. Carneiro, and J. L. R. Pereira, Optimal unbalanced capacitor placement in distribution systems for voltage control and energy losses minimization, *Electric Power Systems Research*, 154, 110–121, 2018, doi:10.1016/j.epsr.2017.08.012.

11. D. Min, J. H. Ryu, and D. Choi, A long-term capacity expansion planning model for an electric power system integrating large-size renewable energy technologies. *Computers & Operations Research*, 96, 2017, doi:10.1016/j.cor.2017.10.006.

12. B. Vyas, M. P. Sharma, and S. Jain, "Feeder Reconfiguration of distribution network using Minimum Power Flow Methodology," in *2015 Annual IEEE India Conference (INDICON)*, New Delhi, India, 2015, pp. 1–6, doi:10.1109/INDICON.2015.7443229.

13. M. Carrión and R. Zárate-Miñano, Operation of renewable-dominated power systems with a significant penetration of plug-in electric vehicles, *Energy*, 90(1), 827–835, 2015, doi:10.1016/j.energy.2015.07.111.

14. N. Sharma, M. P. Sharma, S. Singh, and B. Vyas, "Methodology for valuation of shunt capacitor bank in power grid," in *2019 10th International Conference on Computing, Communication and Networking Technologies (ICCCNT)*, Kanpur, India, 2019, pp. 1–7, doi:10.1109/ICCCNT45670.2019.8944425.

15. S. Prajapat, M. P. Sharma, S. Johri, and B. Vyas, "Loss reduction & reactive power support in Rajasthan power system: A case study," in *2017 Innovations in Power and Advanced Computing Technologies (i-PACT)*, Vellore, 2017, pp. 1–6, doi:10.1109/IPACT.2017.8245057.

16. S. Corsi, "The secondary voltage regulation in Italy," in *2000 Power Engineering Society Summer Meeting (Cat. No.00CH37134)*, Seattle, WA, USA, 2000, vol. 1, pp. 296–304, doi:10.1109/PESS.2000.867599.

17. S. Impram, S. V. Nese, and B. Oral, Challenges of renewable energy penetration on power system flexibility: A survey, *Energy Strategy Reviews*, 31, 100539, 2020, ISSN 2211-467X, doi:10.1016/j.esr.2020.100539.

18. B. Kroposki, Integrating high levels of variable renewable energy into electric power systems, *Journal of Modern Power System and Clean Energy*, 5, 831–837, 2017, doi:10.1007/s40565-017-0339-3.

19. S. Corsi, P. Marannino, N. Losignore, G. Moreschini, and G. Piccini, "Coordination between the reactive power scheduling function and the hierarchical voltage control of the EHV ENEL system," in *IEEE Transactions on Power Systems*, May 1995, vol. 10, no. 2, pp. 686–694, doi:10.1109/59.387904.

20. D. Thukaram and K. Parthasarathy, An expert system for voltage control in Grid Connections. in *Proceedings 1995 International Conference on Energy Management and Power Delivery EMPD'95*, 1995, vol. 1, pp. 364–369, doi:10.1109/EMPD.1995.500754.

21. P. Rani, M. P. Sharma, B. Vyas, and S. Ola, "Loss reduction of Rajasthan power system with Distributed Generation in transmission network," in *2016 IEEE 1st International Conference on Power Electronics, Intelligent Control and Energy Systems (ICPEICES)*, Delhi, 2016, pp. 1–6, doi:10.1109/ICPEICES.2016.7853597.

22. R. Panwar, V. Sharma, M. P. Sharma, and B. Vyas, "Circulating MVAR control in Rajasthan (India) transmission system," in *2016 IEEE 1st International Conference on Power Electronics, Intelligent Control and Energy Systems (ICPEICES)*, Delhi, India, 2016, pp. 1–6, doi:10.1109/ICPEICES.2016.7853696.

23. L. Yan, C. Yongning, T. Haiyan, T. Xinshou, Z. Zhankui, and J. Jianqing, "Common focus and new requirement on technical standards of renewable energy grid integration," *2019 Chinese Automation Congress (CAC)*, Hangzhou, China, 2019, pp. 3719–3723, doi:10.1109/CAC48633.2019.8996943.
24. B. Vyas, M. Gupta, M. P. Sharma, and A. Dandotia, "Sustainable development of agriculture feeders by solar-var incorporation (March 19, 2019)," in *Proceedings of International Conference on Sustainable Computing in Science, Technology and Management (SUSCOM)*, Amity University Rajasthan, Jaipur - India, February 26–28, 2019, doi:10.2139/ssrn.3355181.
25. D. Mitra, M. P. Sharma, T. Suman, and B. Vyas, "Identification & reduction of circulating Mvar loops at high voltage substations," in *2020 21st National Power Systems Conference (NPSC)*, Gandhinagar, India, 2020, pp. 1–6, doi:10.1109/NPSC49263.2020.9331763.
26. B. Vyas, M. K. Gupta, and M. P. Sharma, Distributed volt ampere reactive power compensation of modern power system to control high voltage, *Journal of the Institution of Engineers (India): Series B*, 101, 93–100, 2020, doi:10.1007/s40031-020-00422-3.
27. S. M. Abd-Elazim and E. S. Ali, Load frequency controller design of a two-area system composing of PV grid and thermal generator via firefly algorithm, *Neural Computing and Applications*, 30, 2 (July 2018), 607–616, 2018, doi:10.1007/s00521-016-2668-y.
28. R. K. Agrawal, M. P. Sharma, N. K. Kumawat and B. Vyas, "High Voltage Mitigation of EHV System by Shunt Reactor vs Shunt Capacitor," in *2019 8th International Conference on Power Systems (ICPS)*, 2019, pp. 1–6, doi:10.1109/ICPS48983.2019.9067594.

4 Energy Storing Devices for Sustainable Environment

Radhika G. Deshmukh
Shri Shivaji Science College

CONTENTS

DOI: 10.1201/9781003242277-4

4.1   INTRODUCTION

The demand for energy is increasing rapidly due to the increase in population, eco-nomic development in developing countries, increase in per capita consumption, change in lifestyle, and supply at more remote places as stored energy. The world's primary energy consumption was 149,634 and 157,064 terawatt-hours (TWh) in 2015 and 2018, respectively. The biggest consumers of energy were Asia Pacific and North America, while Africa used the least amount of energy in 2018. For balancing and matching the demand and supply, the storage of energy is a necessity. The present trends indicate that the need for energy storage will increase with high production and demand, necessitating energy storage for many days or weeks or even months in the future. In line with the technological progress made during industrialization, the energy requirements in the industrialized world are increasing. One of the main technical inventions is the discovery of electricity. Heat and electricity production from these sources is able to meet the demand, which reduces additional energy stor-age [1,2]. Our needs have long been met by fossil fuels such as coal, oil, and natural gas. However, these energy sources are limited and have negative environmental effects. Old energy sources are partially replaced by modern energy sources such as nuclear and renewable resources. People's growth, developing countries' economic development, increase in per capita consumption, and lifestyle changes quickly lead to increased energy demand. Energy storage is required to balance the demand and supply. These trends demonstrate that the need for energy storage increases with high production and demand. The requirement for energy storage is estimated to increase threefold, according to the estimates, by 2030. With nuclear accidents and global warming as well as rising prices and small amounts of fossil fuels, the amount of energy obtained from various sources has increased and the percentage of renewable energy sources has increased at the moment [3,4]. There are environ-mentally less harmful and inexhaustible sources of renewable energy, such as sun and wind. However, they are unpredictable and harder to regulate. Therefore, one of the biggest challenges is to tune the available energy with energy demand in time, place, and quantity. The energy storage technology and equipment were reformed and modernized in conjunction with increasing production and demand. Currently, a very high number of energy storage technologies (ESTs) such as electrochemical energy storage (ECES), mechanical energy storage (MES), chemical energy storage (CES), and TES are available [5]. The advantaged technology is hydropower stor-age (mechanical), traditionally container batteries, dry cells, and condensers, and currently provides advantages and disadvantages for all technologies [6–9]. In addi-tion to making changes in a range of old devices, however, a number of new ESTs and systems have been developed for sustainable, renewable energy storage, such as flow batteries, supercondenser systems, flywheel energy storage (FES), and magnetic superconducting magnetic energy storage (SMES). For optimizing renewable energy storage and its use, energy storage research has become essential. Many research studies have proposed ESTs and compared them.

4.2 PROBLEM

The lack of knowledge allows users to understand the advantages and inconveniences of various technologies, but also allows them to decide which storage systems suit the application best. Energy storage is a relatively new subject of research, however. The number of available technologies for energy storage is increasing every day. This chapter compares various ESTs and provides the best energy storage device for various applications [10,11].

4.3 GLOBAL STATUS OF THE CONSUMPTION OF ENERGY

The energy consumption has increased tremendously after the industrial revolutions due to the increase in population, invention of new techniques and machines, economic development, access to remote and far-flanged areas, and big changes in the lifestyle. According to estimates, the energy use had doubled in each decade in the past. Simultaneously, a significant increase also took place in the production of energy, especially electricity. Among the other drivers of increasing demand for energy are selling of the electricity for low cost in GCC and some other countries, wastage due to usage and building designs, and lower efficiency of generation and delivery equipment. Nevertheless, the production could not match the demand in so many developing countries. According to estimates, the world's primary energy consumption in 2015 remained as 146,000 terawatt-hours (TWh), 25 times higher than the year 1800. As the data values are not mostly the same when reported by different sources, in another report, the global energy consumption was 136,129 TWh in 2008 and 161,250 TWh in 2018. There has been a 2.9% increase in consumption in 10 years. While finding scenarios and exploring innovative pathways to 2040 contemplate that the globe will be entering a new energy era promising enough, clean, and sustainable energy for all communities with increasing uses and users. About 10% increase is presumed in the demand of energy by 2040. However, there will be more emphasis on renewable sources considering environmental protection. The global energy consumption by region is given in Figure 4.1.

The energy consumption has increased tremendously after the industrial revolutions due to an increase in population, invention of new techniques and machines, economic development, accessing remote and far flanged areas, and big changes in the lifestyle. According to estimates, energy use was doubling in each decade in earlier times. Simultaneously, a significant increase also took place in the production of energy. While finding scenarios and exploring innovative pathways to 2050, contemplate that the globe will be entering in a new energy era promising enough, clean, and sustainable energy for all communities with increasing uses and users. About 10% increase is presumed in demand of energy by 2050.

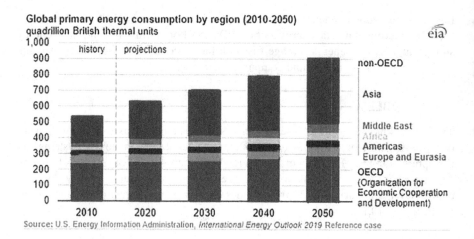

FIGURE 4.1   Global energy consumption by region.

4.4   NEED TO STORE ENERGY

The energy generated from renewable energy sources cannot be stored for later use. The requirement and urgency for energy storage are not uniform and vary. Energy storage can reduce grid energy and improve the functioning of the grid system. A reliable storage of electrical energy increases the safety and economic value and decreases the carbon dioxide emissions. There is a need for special devices and mechanisms to store electricity, which researchers and technologists are improving and innovating. This global energy storage capacity is therefore steadily increasing. The storage capacity has increased by 100% compared to 2017 and 2018, as estimated by the IEA. There are thus growing energy storage equipment and systems. For energy supply in remote areas, the power storage battery is used. This kind of energy was used in vehicles, many of whom now have electricity stored. As a result, the need for increased energy storage will increase significantly because of the continuous supply of renewable resources. Therefore, renewable systems and energy storage devices and systems should in future be combined. The demand for energy does not remain uniform throughout a day and the entire year; rather, it drastically varies within a day and during various seasons of the year. Thus, peak and off-peak demands arise within a day and the seasons due to individual needs and climatic effects. These phenomena necessitate the storage of energy.

4.5   ENERGY STORING TECHNOLOGIES

As the technology developed, the energy storage mechanisms improved over different times. As it produces more than 90% energy, the first battery power storage is still the most used technology. Volta's cell was invented as the primary battery in 1800. The cell of Daniel was transformed from Volta's cell in 1836. In Daniel's cell, two electrolytes were used [12,13]. Then, in 1866 Leclanche cell was formed, which was comprised of a zinc anode and a carbon cathode. In 1948, dry cells were invented. Secondary batteries such as plum acid, Ni–Cd, and Li-ion (Li–O_2 and Li–S) batteries were found in the mid-1980s. In 1991, Li-ion batteries (LIB) were found. The planet has now started using digital techniques, and the power storage equipment is being modernized. The other widely used power storage devices are the condensers and supercondensers. This type of condensers can supplement or substitute electricity storage batteries. Nanostructure carbons are highly porous and have a vast surface area capable of maximizing electrode performance by functional groups. Electricity can be stored via different systems, and a transmission system for the storage purposes of compressed-air and pumped hydroelectric energy is designed. The collection of all electricity storage methods and systems associated with the grid system in large quantities is referred to as grid energy storage or large-scale energy devices. The technology and devices for energy storage often are classified into different categories. The mechanisms or storage devices may be mechanical (solar fuel cells and pumped hydroelectric energy storage), electrical (conventional rechargeable batteries and flow pumps), chemical (hydrogen storage), and electronic (capacitors, supercapacitors, and SMES).

4.6 CLASSIFICATION OF ENERGY STORAGE

There are many possible techniques for energy storage: mechanical, electrical, energy predictable (for example, the delivery of electrical potential, chemical energy, and thermal energy) [14–18]. These have all been leading to the birth of the techniques that the storage technologies which vary considerably as a function of the applications and needs, will obviously be of different types. The technologies are many, but a comparative study is rendered difficult by the fact that, among others, their levels of development vary with different types.

The storage devices can be classified as follows:

1. **Technology Type**: Mechanical, electrical, electrochemical, and thermal.
2. **Power and Energy Rating**: Large-scale PHES, CAES, fuel cells, TES, and flow batteries; small-scale FES, BES, Micro-Compressed 0 200 400 600 OO 1000 1200 1400 air energy storage (MCAES), supercapacitors, and SMES.
3. **Application Type**: Power quality and reliability, self-discharge, and internal resistance power network.

   According to their applications, the storage techniques can be divided into four categories:

   i.   Low-power application in isolated areas, essentially to feed transducers and emergency terminals.
   ii. Medium-power application in isolated areas (individual electrical systems and town supply).
   iii. Network connection application with peak leveling and discharge.
   iv. Power quality control applications.

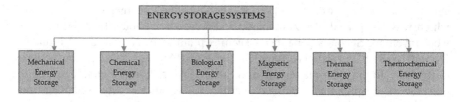

FIGURE 4.2 Schematic flowchart of classification of energy storing devices.

The first two categories are for small-scale systems, where energy could be stored as kinetic energy, chemical energy (accumulators and flow batteries), compressed air, hydrogen (fuel cells), or electricity in supercapacitors. The chart of classification of energy storage devices is given in Figure 4.2.

4.6.1 Electrochemical Energy Storage (ECES)

These are electrochemical devices that convert the stored energy into electricity either primarily or secondarily. Both flow batteries and solid-state batteries of this type are quite small and portable. The electrical energy storages are lithium-ion battery (LIB) [15–17]. The electronic energy storages are sodium–sulfur battery, lead–acid battery, and redox flow batteries.

4.6.1.1 Lithium-Ion Batteries (LIB)

LIB are made of two low-density lithium electrodes and have a large standard electrical potential, making them the main electronic energy handler. They are a low-weight and high-voltage battery without memory and with low self-loads and IOT. LIB are used in mobiles and electrical vehicles, but for large-scale grid storage, they are too expensive. The research is, however, comprehensive. LIB have an extensive impact on the depletion of metals and can therefore cause significant environmental, social, and health impacts due to its toxicity. The toxicity of LIB is, for sure, lower than that of many other batteries [19,20]. A schematic diagram of a Li-ion battery is given in Figure 4.3.

4.6.1.2 Lithium–O₂ Batteries

These batteries possess high energy density as compared to Li-ion batteries. This high value of energy density is suitable for renewable energy storage applications. The anode of this battery is made of lithium metal and oxygen works as the cathode; both are available in the environment and the battery's overall weight is reduced. A typical $Li–O_2$ battery consists of an anode made of lithium metal, a cathode made of porous carbon, and conductive electrolyte of Li-ion.

4.6.1.3 Lithium Cobalt Oxide Batteries

A cobalt oxide cathode and a graphite carbon anode are used in these lithium cobalt oxide ($LiCoO_2$) batteries. The anode (graphite) and the cathode ($LiCoO_2$) are

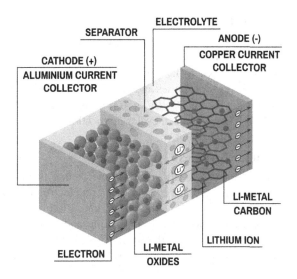

FIGURE 4.3 Schematic diagram of description of Li-ion battery.

separated by a non-aqueous liquid electrolyte. As these batteries have a high specific energy, these can be used in laptops and digital cameras. The energy density of these batteries is from 150 to 200 Wh/kg. The performance is good, but they have some drawbacks. These batteries have a short life span and low power capability. These batteries may damage due to overheating.

4.6.1.4 Lithium Manganese Oxide Batteries

These are three-dimensional spinel structure batteries. The cathode of these batteries is made of lithium manganese oxide material. As these batteries have a high thermal stability, they are highly recommended for use in high-load applications such as power tools and electric vehicles. These batteries have a low life span, and the energy density is between 100 and 150 Wh/kg.

4.6.1.5 Lithium Nickel Manganese Cobalt Oxide Batteries

These batteries have flexible power storage, are cost-effective, and are used in various applications in electrical and medical devices. These are the combination of nickel, manganese, and cobalt. These can be operated in the range of 3–4.2 V like Li-ion batteries. The battery life reaches up to 2,000 cycles, and the minimum and maximum energy densities are 150 and 220 kWh/kg.

4.6.1.6 Lithium Iron Phosphate Batteries

These batteries have good thermal stability, lower stress at high voltage even for longer time, high tolerance at full charge conditions, high current rating, and long cycle life. In these batteries, phosphate is used as the cathode, which has low resistance and electrochemical performance. Due to self-discharging, they have a low specific energy density of about 90–120 kWh/kg. These batteries can reach up to about 2,000 cycles.

4.6.1.7 Lithium-Titanate Batteries

These batteries use anodes made of titanate material, and the material for cathode used is as in lithium nickel manganese cobalt oxide batteries. They perform well at high temperatures. When faced to fast charging or charging at low temperatures. These are uninterruptible power supplies and powertrains, but its low specific energy density is a disadvantage, which is 50 Wh/kg. It is of high cost with a typical operating range of 1.8–2.85 V. They can reach cycles up to 7,000.

4.6.1.8 Sodium–Sulfur Batteries

Sodium battery is made up of two sulfur molten electrodes as a positive electrode, and sodium molten as a solid sodium electrode and an electrolyte as a negative element. A SSB has low density, high specific power, long lifetime, and high load efficiency. The SSB was used to facilitate load leveling and renewable integration in Reunion Island, an island of France in the Indian Ocean. The SSB has a capacity of 1 MW and can deliver power to an average of 2,000 households. A working sodium–sulfur battery is shown in Figure 4.4.

4.6.1.9 Lead–Acid Batteries (LABs)

Lead–acid batteries were the first battery to be recharged. They usually include two cathodes (PbO_2) and a lead (Pb) electrode immersed in sulfuric acid (H_2SO_4). These have low density, low cost, and increased energy density. LABs emit large amounts of lead, which is a toxic metal with potential health risks and bioaccumulates globally. The LAB is also used to store backup power. LABs are often repeatedly recycled and are currently the leading consumer product. LABs have a short life and are a mature technology to store highest current levels (Figure 4.5).

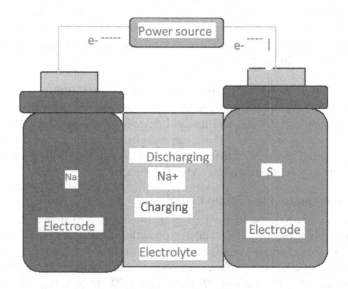

FIGURE 4.4 Schematic diagram of sodium–sulfur battery.

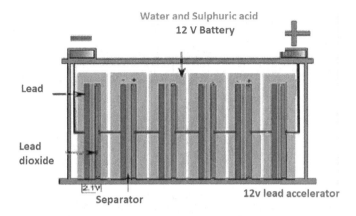

FIGURE 4.5 Schematic diagram of description of lead–acid battery.

4.6.1.10 Aluminum-Ion Batteries

Aluminum-ion batteries in which aluminum is at anode function as the electrochemical disposition and anions in the graphite cathode. These batteries have a power density of 3,000 W/kg and an energy density of 40 Wh/kg. These batteries are safe, are cost-effective, and have a high power density. Furthermore, these batteries are under research for ultrafast charge and discharge along with grid energy storage applications.

4.6.1.11 Copper Zinc Batteries

Copper zinc batteries use the aqueous electrolyte by "Cumulus Energy Storage," which is based on the processes used in metal refining. This project of developing copper zinc batteries is focused on cost-effectiveness and safety of the systems. Furthermore, the design capacity of these batteries ranges from 1 to 100 MWh.

4.6.1.12 Redox Flow Batteries

These batteries generate electricity through the potential difference between the tanks. The solution contained by both the tanks become the same containing both positive and negative ions. Vanadiumpolyamide, vanadiumhydrogenobromine, polysulfide bromide, and iron chromium are the common materials being used in these batteries. There are one or more electroactive components used as solid layer in hybrid flow batteries. Therefore, there is no change of phase of two liquid electrolytes with metal ions dissolved in the fluid electrolyte [21–22]. An ion-selective membrane is used to separate the negative and positive redox ions. These batteries have high efficiency, long cycle life, flexible design, and high energy storage capacity, whereas their density is low in comparison with others.

The construction of a redox flow battery is shown in Figure 4.6.

4.6.1.13 Vanadium-Based Flow Batteries

In these batteries, sulfuric acid is used as the electrolyte, which stores electrical energy in the chemical energy form. Ion exchange membranes, electrolytes, and

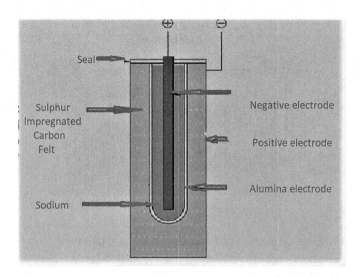

FIGURE 4.6   Redox flow battery.

electrodes are the focus of research in these batteries. These batteries have high power, long life, high efficiency, high safety, and high capacity. However, there are some disadvantages, so these are not commercialized/industrialized.

4.6.1.14   Metal–Air Batteries

Metal–air batteries possess an anode of pure metal, a cathode of air, an electrolyte, and a separator, which is an insulator and allows the transfer of ions only. These batteries possess much higher specific energy capacity.

4.6.2   MECHANICAL STORAGE

Mechanical storage of energy is used to store and transfer energy to longer distances. Compressed-air and pumped hydroelectric energy storages are the latest technology.

4.6.2.1   Hydroelectric Energy Storage

At hydropower plants, the kinetic energy of flowing and falling water is converted into electricity. The generation of electricity is from mechanical energy. Pumped hydroelectric energy storage (PHES) is a process of storing and generating energy using two water reservoirs or basins located at different elevations by pumping water from a lower reservoir to a higher reservoir, and in CAES, the pumping of air into a large reservoir is based on a high-pressure external power supply.

- **Compressed-Air Energy Storage (CAES):** The pumping of air into a large reservoir is based on a high-pressure external power supply. The CAES has a high capacity for energy storage. It can be started quickly with long storage capacity (usually 10 minutes). CAES systems need certain geological requirements for their installation (e.g., salt caverns). These conditions are worldwide limited. However, several concepts for air tanks are being developed so that the constraints can be overcome.

- **Pumped Hydroelectric Energy Storage (PHES)** [23–25]: In times of low demand in pumped hydroelectric energy storage systems, water is pumped into an upper reservoir from a lower reservoir. PHES has a low density, large storage capacity, and about 70%–85% efficiency. The technologies used are state of the art. The technology is favored by high demand, because it provides high capacity and power. PHES has a short start-up time, but is restricted geographically. Usually, PHES is used to store energy between 2 and 8 hours as a medium-term storage system. PHES systems are coupled with wind and solar photovoltaic power. The PHES is the most mature large-scale EST available across the world. It provides an electrical storage capacity of about 99%, with a total installed capacity of more than 120 GW, and contributes to 3% power generation. The PHES offers a very low energy density, almost zero self-discharge, reasonable price per unit stored energy, and high round-trip efficiency. The major disadvantages of the PHES plant are associated with the need for sufficient water availability and appropriate geographical morphology.

4.6.2.2 Flywheel

Flywheel is a mechanical rotating device for storing rotating energy. In its center, a flywheel contains a spinning mass, driven by an engine; when it needs energy, the spinning force drives a tool to supply electricity almost like a turbine, which it slows down. By using a motor, the flywheel is recharged to increase its speed again. A flywheel can capture energy over a period of time from intermittent energy sources and supply the grid with endless power supply. Flywheels are ready to respond immediately to grid signals, providing frequency regulation and improvement in electricity quality. More advanced flywheel designs consist of materials of carbon fiber, stored in vacuum, which allow magnetic levitation an not tradition roller bearings to be rotated up to speeds of 60,000 RPM. Flywheels are further developed by using four key features:

- Rotative weights consisting of fiberglass, or high-speed polymer resin; the FESS achieves an attractive energy density, high efficiency, and low standby efficiency.
- Mass operating in vacuum to dampen aerodynamic drag.
- High-frequency rotating weight.

Air or magnetic suppression technologies. For low-power applications, these require many cycles. These have a number of benefits.

4.6.3 CHEMICAL STORAGE

The energy storage systems that store electricity and heat in chemical bonds for future energy supply are developed by chemical reactions. Energy storage is done by using the chemical technology. The chemical ESTs are (i) hydrogen fuel and (ii) methane.

4.6.3.1 Hydrogen

The efficiency of hydrogen fuel is lower, but carbon emissions are negligible. It is easier to use from environmental perspective. Long-term wind and solar storage technology is a deficit and can even balance seasonal differences. After storage, the hydrogen in an internal combustion engine or a cell is often converted back to electricity or heat. The technology can be used as a carburizer in portable vehicles such as rocket units.

4.6.3.2 Methane

In a multi-step process, the hydrogen reacts with carbon dioxide during electrolysis of water followed by methalization reaction in which methane and water are produced (CO_2). The generated methane can then be stored in gas cavities or ideally in a gas network. Methane can all be recycled from the existing gas grid; it is favorable for both social and environmental reasons. Methane can actually be stored in the existing gas grid.

4.6.4 THERMAL ENERGY STORAGE

Heat or cooling storage medium can be used to store thermal energy. The stored energy can then be used to generate electricity in cooling and heating applications. The three main ways in which materials can store thermal energy are sensitive heat, heat transformation, and chemical reactions. Three technologies exist for thermal energy storage. Primary energy demand is mainly satisfied by fossil fuels, which has pushed them to the brink of depletion. To decelerate the fast depletion of mineral sources of fuels, it is crucial to seek alternative resources that are renewable and sustainable in nature. Significant contribution to pollution reduction can also be observed, such as reduction in carbon dioxide (CO_2) emission through promotion of renewable energy usage. Solar thermal systems have shown tremendous growth in terms of efficiencies, but lack the backup to be on the forefront of energy generation despite being a relatively mature technology. TES systems are commonly employed in construction as well as industrial applications due to their advantages such as improved overall efficiency and better reliability.

The TES can be categorized into three forms:

1. Sensible heat storage (SHS)
2. Latent heat storage (LHS)
3. Thermochemical heat storage (THS).

4.6.4.1 Sensible Heat Storage (SHS)

It is an advanced technology involving heat storage through cooling or heating of a solid storage device or liquid. Sensible heat storage is a system where the energy is stored by increasing or decreasing the temperature of the storage material. This storage medium is available as either solid or liquid. Water is one of the most commonly used media as it is the cheapest. The sensible heat storage works based on heat capacity and corresponding temperature variation of the respective storage material when the charge and discharge phases take place. For temperatures greater than 100 °C, containment media such as liquid metals, oils, and molten salts are much preferred.

4.6.4.2 Latent Heat Storage

The latent heat storage (LHS) includes a material known as a phase change material (PCM). The PCM changes its phase at a particular temperature called the phase change temperature (PCT). The LHS systems have a low thermal conductivity, have a high density of storage, and are corrosive in nature. The LHS systems commonly employ phase change materials (PCMs) for energy containment. PCMs have the potential for energy absorption as well as energy release, which involves physical state change. In a LHS system, heat containment process takes place during the phase change that occurs over a relatively stable temperature and corresponds proportionally to the fusion heat of substance. The PCMs portray themselves as materials that display elevated energy containment as well as maintain a uniform temperature throughout the process.

4.6.4.3 Thermochemical Energy Storage

Thermochemical energy storage (TCES) stores thermal energy through chemical reactions. In this storage, two or more components combined are then separately stored in a chemical compound that breaks down through heat and split components. During high demand periods, the parts are reassembled in a chemical compound and heat is released. The heat that the reaction releases is the storage capacity. TCES makes it possible to bridge long periods between demand and supply. This makes TCES particularly well suited for power generation. The efficiency of this technology ranges from 75% to nearly 100%, and TCES materials in all storage media are among the densest. However, TCES-based storage technology is mainly being developed [26–31]. The graphical review and comparison of ESTs on the basis of energy and power is given in Figure 4.7.

The comparative chart of different energy storing technologies is discussed here from Table 4.1. We compare all ESTs on the basis of advantages, disadvantages, power application, and energy application. It is observed that almost all ESTs have a high capacity and high energy density. Flywheels and SMES have high power. All technologies are feasible, except for metal–air, pump storage, and CAES. EC capacitors have long life cycle as compared to other technologies. All ESTs have some disadvantages. Flywheels and EC capacitors are the best (Figures 4.8–4.10).

All energy storing technologies are compared on the basis of their technological performance and characteristics such as power range, energy density, power density, efficiency, discharge time, response time, lifetime, self-discharge, and technology maturity level. Technical characteristics are summarized in Tables 4.2 and 4.3. Energy and power densities of all selected ESTs are compared; electrochemical energy storage systems (Pb–A, Ni–Cd, Na–S, and Li-ion) have higher energy density than others. On the other hand, the power density of FES, SMES, and Strongly Correlated Electron System (SCES) is higher than that of other types of ESTs. Among electrochemical energy storage systems, Li-ion and Na–S have higher energy density than other selected rechargeable batteries. Among all selected ESTs, PHES and CAES have a higher power range and longer discharge time than others. SMES, FES,

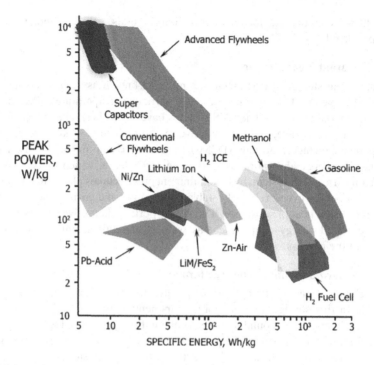

FIGURE 4.7   Comparison of ESTs' power and specific energy.

and SCES have a low power range and a very short discharge time. PHES, CAES, and VRFB have a very small daily self-discharge ratio compared to other types of ESTs. Among the electrochemical energy storage systems, Na–S systems have a high self-discharge rate per day. Similarly, the electrical energy storage systems SMES and SCES have a high self-discharge rate per day. Compared to all selected ESTs, PHES has the largest life span of 50 years, while electrochemical energy storage systems fall within the range of 7.5–15 years; PHES, Pb–A, and Ni–Cd batteries are the most mature and fully commercialized ESTs compared to others. FES, SMES, SCES, and Li-ion batteries have a very high cycle efficiency of above 90%, which are the top among ESTs. Similarly, PHES, CAES, and electrochemical energy storage systems (except for Li-ion) have a cycle efficiency in the range of 74.5%–90%. FES, SMES, and SCES offer very fast response time in the order of milliseconds; electrochemical energy storage systems in the order of seconds, and PHES and CAES in the order of minutes. The capital cost per kWh of PHES and CAES is in the low range category compared to other types of ESTs. On the other hand, FES, SMES, and SCES have a higher capital sustainability. CAES has the lowest energy capital cost compared to all other technologies. Among the electrochemical energy storage systems, Ni–Cd, Li-ion, and Na–S have a higher capital cost per kWh than other rechargeable batteries. PHES, CAES, electrochemical energy storage systems, and SMES have a high negative impact on the environment, whereas FES and SCES have very low impacts on the environment [32–39].

TABLE 4.1

Comparative Chart of Different Energy Storage Technologies

Technology	Advantages	Disadvantages	Power Application	Energy Application
CAES	High capacity and low cost	Special site requirements; need gas fuel	Not feasible or economical	Fully capable and reasonable
Flow battery	High capacity; independent power and energy rating	Low energy density	Reasonable for this application	Fully capable and reasonable
Metal–air	High energy density	Difficult to charge	Not feasible or economical	Fully capable and reasonable
Na–S	High power and energy densities and high efficiency	Low production cost; safe	Fully capable and reasonable	Fully capable and reasonable
Li-ion	High power and energy densities and high efficiency	High production cost; special circuit required for charging	Fully capable and reasonable	Feasible, but not quite practical or economical
Ni–Cd	High power and energy densities and high efficiency	Higher self-discharge rate	Fully capable and reasonable	Reasonable for this application
Pumped storage	High capacity; low cost	Special site requirements	Not feasible or economical	Fully capable and reasonable
Lead–acid	Low capital cost	Limited life cycles	Fully capable and reasonable	Feasible, but not quite practical or economical
Flywheel	High power	Low energy density	Fully capable and reasonable	Feasible, but not quite practical or economical
SMES	High power	High production cost; low energy density	Fully capable and reasonable	Not feasible or economical
EC capacitors	Long life cycle; high efficiency	Low energy density	Fully capable and reasonable	Reasonable for this application

4.7 DISCUSSION AND ANALYSIS

The outcomes of the present study might help in understanding the need for energy storing devices and the need for performing a comparative study on them. The first one is the objective of storing energy; some of the factors considered to decide the feasibility of a storage system or device are storage capacity; easy load leveling; time required for storage and regeneration; lifetime of the device; and quality, consistency, and

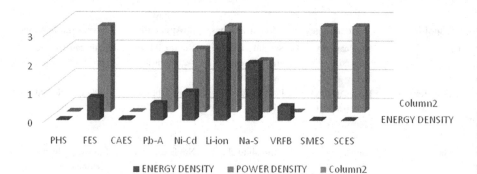

FIGURE 4.8 Comparative study of ESTs on the basis of energy and power densities.

TABLE 4.2
Technical Characteristics of Select ESTs

ESTs	Energy Density (Wh/l)	Power Density (W/l)	Round-Trip Efficiency (%)	Power Range (MW)
PHES	0.5–1.5	0.5–1.5	75–85	10–5,000
			65–87	
			70–85	
FES	20–80	1,000–2,000	93–95	0.1–20
			90–95	< 0.25
			90–93	0–0.25
				0.01–0.25
CAES	3–6	0.5–2	70–89	5–1,000
			50–89	5–300
			70–79	
Pb–A	50–80	10–400	70–90	0–40
			75–80	0–20
Na–S	150–230	150–230	85–90	0.05–34
			80–90	0.05–8
Li-ion	200–500	500–2,000	85–90	0–100
			90–97	0–1
Ni–Cd	60–150	150–300	60–65	0–40
			85–90	
VRFB	16–33	0.5–2	85–90	**0.3-3**
	20–70		75–82	
SCES	2.5–15	500–5,000	90–95	0–0.3
			95–98	
SMES	0.2–2.5	1,000–4,000	95–98	0.1–10

TABLE 4.3
Technical Characteristics of Select ESTs

ESTs	Discharge Time (ms-hr)	Response Time (ms-hr)	Lifetime (yr)	Daily Self-Discharge (%) Technology	Maturity
PHES	1–24 hr+	sec-min, min 1–2 min	**40–60**	Very small	Very mature/fully commercialized
FES	ms–15 min	< 4 ms-sec, sec	15 +, 15	100 24–100	Mature/being commercialized
CAES	1–24 hr+	1–15 min 1–2 min	20–40	small 0.00	Proven/being commercialized
Pb–A	sec-hr	5–10 ms	3–15, 5–15	0.1–0.3 ,0.033–1.10	Very mature/fully commercialized
Ni–Cd	sec-hr	20 ms-sec, sec	10–20	0.2–0.6, 0.07–0.71	Very mature/fully commercialized
Na–S	sec-hr	1 ms, sec	10–15	20	Proven/being commercialized
Li-ion	min–hr	20 ms-s	5–15	0.1–0.3, 0.03–0.33	Proven/being commercialized
VRFB	sec-10 hrs	Sec	5–10	Small	Proven/being commercialized
SCES	ms–hr	8 ms	20 +	20–40 0.46–40	Proven/being commercialized
SMES	ms–8 sec	< 100 ms	20 +	10–15 1–15	Proven/being commercialized

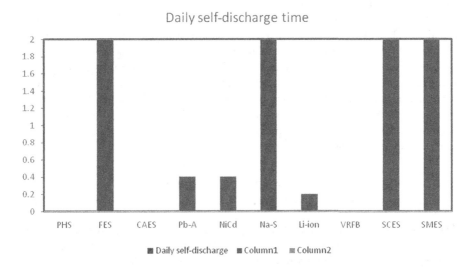

FIGURE 4.9 Comparative study of ESTs on the basis of daily self-discharge time.

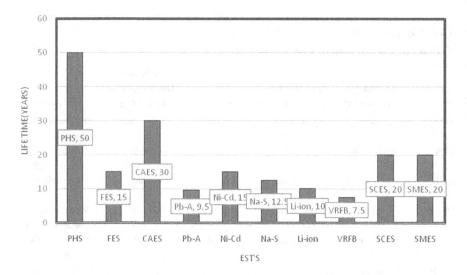

FIGURE 4.10 Comparative study of ESTs on the basis of life time (years).

reliability of the discharged energy. Moreover, cost comparison is highly important as well. The environmental concerns are becoming very important due to the complaints and protests of people, which must be considered while making an implementation decision. Batteries such as lead–acid, Li-ion (Li–O$_2$ and Li–S), and flow batteries have very high capacities and can supply energy even to run heavy vehicles and grids of electricity networks. These can help to increase storage capacities and improve other characteristics of devices. Although batteries are mostly of low cost, their shorter life-cycle and consequent frequent replacement make these costly. CAES stores energy by compressing air and is considered one of the most efficient and economically attractive systems that enable successful load management with large electrical grids. It causes low environmental impacts and has a longer life cycle, but its operational costs are high. PHES uses the potential energy of water transferred between two reservoirs located at different altitudes. It is also a mature technology, but it is expensive in terms of installation cost and require suitable sites for the construction of reservoirs. SMES is the best suitable device to provide constant and instant power supply as well as to regulate grid stability with very high-power output within a short time, and it can provide power with quality to the consumers, although the systems are costly. Hydrogen is an immature technology, but can attract huge interest in future if it is developed in terms of aspects such as generation, storage, and utilization [40–49].

4.8 CHALLENGES AND PROSPECTS OF ENERGY STORAGE TECHNOLOGIES

The challenges and innovations of energy storage devices and systems are commercial, widespread, and long-term adaptations of recent inventions. A few constraints and challenges are faced globally when energy storage devices are used, and storage systems are in operation for storing the surplus of generated energy. A lot of money is also incurred in the implementation, running, and replacement of the energy storage systems, which in certain cases is quite high. Some energy storage devices can cause environmental problems, so research is required to develop devices that not only have higher efficiencies, but also are cheaper and have minimum environmental problems, especially during the disposal of used devices after completing the life cycle. Standards and regulations must be developed to assess the environmental impacts of various devices and systems. The current energy production is dominated by generation from fossil fuels, which is not only costly, but also non-renewable; therefore, it cannot be sustained indefinitely. The integration of renewable energy sources and energy storage systems to minimize the share of fossil fuel plants is increasing. Hence, technologies to produce energy renewable sources such as wind and solar must be developed. An awareness of energy storage should be created among all the stakeholders for the development and installation of modern and more efficient devices of energy storage. Contribution of the present study for decision making the planners, policy makers, and the practitioners often face problems to select the most appropriate device or the combination of two or more devices/systems to store energy for the grids or static forms. Therefore, they need some broad guidelines [50–58].

4.9   CONCLUSIONS

The major theme and need of storing energy are ensuring its availability when direct sources cannot be captured energy at almost fixed rates while the demands are fluctuating simultaneously. Thus, energy storage makes it possible to supply energy at peak times and to store it at off-peak times. The device used for storing energy in the past is the battery, which has been changing so much and appearing in so many forms. Many modifications have occurred in the devices of energy storage. At present, batteries such as lead–acid, Ni–Cd, Li-ion (Li–O_2 and Li–S), NiMH (nickel–metal hydride), and finally flow batteries are in use. The batteries that are in used today have very high capacities and can supply energy even to run heavy vehicles and grids of electricity networks. Capacitors and supercapacitors are also used to store electrical energy for running utility devices. More advanced mechanisms and systems of storing energy are superconducting magnetic energy storage (SMES), flywheel energy storage (FES), pumped hydroelectric energy storage (PHES), thermal energy storage (TES), compressed-air energy storage (CAES), and hybrid energy storage (HES). Each of these systems/technologies has advantages and constraints, but these can be used to match the requirements of the location and the delivery network as well as the demand of customers. Various types and sizes of batteries are required for storing static energy to run vehicles, machines and equipment, and entertainment and communication devices [59–64]. The storing techniques and devices can also affect the environment. There are some constraints and challenges during the processes of energy storage. None of the devices and systems returns 100% of the stored energy, which means that there is wastage (10%–30%). Research must be conducted, and devices with higher efficiencies should be developed. A few building codes should be implemented. Standards must be developed to assess the environmental impacts of various devices and systems, and regulations must be implemented to control these. Efforts should be made to increase the energy production from renewable resources for decreasing emissions and environmental impacts. The study may lead to the evaluation of the most important technique for a given application, but other aspects, including environmental, social, and ethical, need to be taken into account in order to reach a final decision. The results of this study can therefore be helpful for various stakeholders in energy sector and for creating a sustainable environment by the use of pollution-free energy storing devices.

NOMENCLATURE

CAES: Compressed-air energy storage
ECES: Electrochemical energy
ESTs: Energy storage technologies
FES: Fleet energy storage
FES: Flywheel energy storage
HES: Hybrid energy storage
LHS: Latent heat storage
MES: Mechanical energy storage
PHES: Pumped hydroelectric energy storage
SHS: Sensible heat storage
SMES: Superconducting magnetic energy storage
SSB: Sodium–sulfur battery
TES: Thermal energy storage
THS: Thermochemical heat storage
Twh: Terawatt-hours
VRFB: Vanadium redox flow battery

REFERENCES

1. Gustavssna, J. (2016). Energy storage comparison technology, *Unpublished bachelor thesis*, KTH School of Industrial Engineering and Management Energy Technology, Stockholm.
2. Kyriakopoulos, G. L., Arabatzis, G. (2015). Electrical energy storage systems in electricity generation: Energy policies innovative technologies, and regulatory regimes. *Renewable and Sustainable Energy Reviews*, 56, 1044–1067.
3. World nuclear association. (2016). World nuclear association. http://www.world-nuclear.org/information-library/countryprofiles/countries-a-f/france.aspx.
4. International Energy Agency. (2016). International Energy Agency. http://energyatlas.iea.org/?subject=-1076250891.
5. Shaqsi, A.Z.A., Sopian, K., Al-Hinai, A. (2020). Review of energy storage services, applications, limitations, and benefits. Energy Reports.
6. Patel, K., Arora, K., Kaur, P. (2019). Power monitoring system in solar power plant using LabVIEW. *In 2019 2nd International Conference on Intelligent Computing, Instrumentation and Control Technologies (ICICICT)*, pp. 1011–1015. doi: 10.1109/ICICICT46008.2019.8993249.
7. IRENA. (2020). *Renewable Energy Capacity Highlights*. New York: Irena.
8. Das, C.K., Bass, O., Kothapalli, G., Mahmoud, T.S., Habibi, D. (2018). Overview of energy storage systems in distribution networks: Placement, sizing, operation, and power quality. *Renewable and Sustainable Energy Reviews*, 91, 1205–1230.

9. Palizban, O.; Kauhaniemi, K. (2016). Energy storage systems in modern grids—Matrix of technologies and applications. *Journal of Energy Storage*, 6, 248–259.

10. Li, G., Zheng, X. (2016). Thermal energy storage system integration forms for a sustainable future. *Renewable and Sustainable Energy Reviews*, 62, 736–757.

11. Tomazic, G., Skyllas-Kazacos, M. (2014). Redox flow battery. In *Electrochemical Energy Storage for Renewable Sources and Grid Balancing*, vol. 1, pp. 309–336. Durham: Elsevier

12. Prashar, D., Arora, K. (2020). Design of two area load frequency control power system under unilateral contract with the help of conventional controller. *IJICTDC*, 5, 22–27.

13. Arora, K., Kumar, A., Kamboj, V.K., Prashar, D., Jha, S., Shrestha, B., Joshi, G.P. (2020). Optimization Methodologies and testing on standard benchmark functions of load frequency control for interconnected multi area power system in smart grids. *Mathematics*, 8, 980.

14. Arora, K., Kumar, A., Kamboj, V.K., Prashar, D., Shrestha, B., Joshi, G.P. (2021). Impact of renewable energy sources into multi area multi-source load frequency control of interrelated power system. *Mathematics*, 9, 186.

15. Zhao, H., Wu, Q., Hu, S., Xu, H., Rasmussen, C.N. (2015). Review of energy storage system for wind power integration support. *Applied Energy*, 137, 545–553.

16. Behabtu, H.A., Messagie, M., Coosemans, T., Berecibar, M., Anlay Fante, K., Kebede, A.A., Mierlo, J.V. (2020). A review of energy storage technologies' application potentials in renewable energy sources grid integration. *Sustainability*, 12, 10511.

17. Al Zohbi, G., Hendrick, P., Renier, C., Bouillard, P. (2016). The contribution of wind-hydro pumped storage systems in meeting Lebanon's electricity demand. *International Journal of Hydrogen Energy*, 41(17), 6996e7004.

18. Angell, M., Pan, C.-J., Rong, Y., Yuan, C., Lin, M.-C., Hwang, B.-J., Dai, H. (2017). High Coulombic efficiency aluminum-ion battery using an AlCl3-urea ionic liquid analog electrolyte. *Proceedings of the National Academy of Sciences*, 114(5), 834. doi:10.1073/pnas.1619795114.

19. Anne-Chloe´ Devic, M. I., Sua´rez, E., Ferna´ndez, V., Bax, L. (2018). Battery energy storage. SUSCHEM European technology platform for sustainable chemistry. Retrieved on 02.05.2019, www.suschem.org/files/library/Suschem_energy_storage_final_preview.pdf.

20. Ayyappan, S., Mayilsamy, K., Sreenarayanan, V.V. (2016). Performance improvement studies in a solar greenhouse drier using sensible heat storage materials. *Heat and Mass Transfer*, 52(3), 459e467. doi:10.1007/s00231-015-1568-5.

21. Bamgbopa, M.O., Almheiri, S., Sun, H. (2017). Prospects of recently developed membraneless cell designs for redox flow batteries. *Renewable and Sustainable Energy Reviews*, 70, 506e518. doi:10.1016/j.rser.2016.11.234.

22. Battke, B., Schmidt, T.S., Grosspietsch, D., Hoffmann, V.H. (2013). A review and probabilistic model of lifecycle costs of stationary batteries in multiple applications. *Renewable and Sustainable Energy Reviews*, 25, 240e250.

23. Benato, A., Stoppato, A. (2018). Pumped thermal electricity storage: A technology overview. *Thermal Science and Engineering Progress*, 6, 301e315.

24. Hussain, F., Rahman, M.Z., Sivasengaran, A.N., Hasanuzzaman, M. (2020). Energy storage technologies, in Energy for Sustainable Development, pp. 125–165. Academic Press

25. Leadbetter, J., Swan, L.G. (2012). Selection of battery technology to support grid-integrated renewable electricity. *Journal of Power Sources*, 216, 376–386.

26. Johal, H., Manz, D., O'Brien, K., Kern, J. (2011). Grid integration of energy storage. In *Proceedings of the IEEE Power and Energy Society General Meeting*, Detroit, USA; pp. 1–2.

27. Sabihuddin, S., Kiprakis, A.E., Mueller, M. (2015). A numerical and graphical review of energy storage technologies. *Energies*, 8, 172–216.

28. Denholm, P., Ela, E., Kirby, B., Milligan, M. (2011). The role of energy storage with renewable electricity generation. In *Energy Storage: Issues and Applications*, Elsevier, pp. 1–58. Oak Ridge, TN: NREL.
29. Luo, X., Wang, J., Dooner, M., Clarke, J. (2015). Overview of current development in electrical energy storage technologies and the application potential in power system operation. *Applied Energy*, 137, 511–536.
30. Rohit, A.K., Rangnekar, S. (2017). An overview of energy storage and its importance in Indian renewable energy sector: Part II—Energy storage applications, benefits and market potential. *Journal of Energy Storage*, 13, 447–456.
31. Chen, H., Cong, T.N., Yang, W., Tan, C., Li, Y., Ding, Y. (2009). Progress in electrical energy storage system: A critical review. *Progress in Natural Science*, 19, 291–312.
32. Nadeem, F., Hussain, S.M.S., Tiwari, P.K., Goswami, A.K., Ustun, T.S. (2019). Comparative review of energy storage systems, their roles, and impacts on future power systems. *IEEE Access*, 7, 4555–4585.
33. Ibrahim, H., Ilinca, A., Perron, J. (2008). Energy storage systems-Characteristics and comparisons. *Renewable and Sustainable Energy Reviews*, 12, 1221–1250.
34. Díaz-González, F., Sumper, A., Gomis-Bellmunt, O., Villafáfila-Robles, R. (2017). A review of energy storage technologies for wind power applications. *Renewable and Sustainable Energy Reviews C*, 16, 2154–2171.
35. Arora, K., Singh, B., (2012). Load frequency control of Interconnected hydro – Thermal system with conventional and expert controllers. Buletin *Teknik Electro dan Informatika*, 1 (4), 255–262.
36. Maisanam, A.K.S., Biswas, A., Sharma, K.K. (2020). An innovative framework for electrical energy storage system selection for remote area electrification with renewable energy system: Case of a remote village in India. *Journal of Renewable and Sustainable Energy*, 12, 024101.
37. Koohi-Fayegh, S., Rosen, M.A. (2020). A review of energy storage types, applications and recent developments. *Journal of Energy Storage*, 27, 101047.
38. Acar, C. (2018). A comprehensive evaluation of energy storage options for better sustainability. *International Journal of Energy Research*, 42, 3732–3746.
39. Liu, Y., Du, J.L. (2020). A multi criteria decision support framework for renewable energy storage technology selection. *Journal of Cleaner Production*, 277, 122183.
40. Rahman, M.M., Oni, A.O., Gemechu, E.; Kumar, A. (2020). Assessment of energy storage technologies: A review. *Energy Conversion and Management*, 223, 113295.
41. Thomas, D., D'Hoop, G., Deblecker, O., Genikomsakis, K.N., Ioakimidis, C.S. (2020). An integrated tool for optimal energy scheduling and power quality improvement of a microgrid under multiple demand response schemes. *Applied Energy*, 260, 114314.
42. Thomas, D., Kazempour, J., Papakonstantinou, A., Pinson, P., Deblecker, O., Ioakimidis, C.S. (2020). A local market mechanism for physical storage rights. *IEEE Transactions on Power Systems*, 35, 3087–3099.
43. Kebede, A.A., Berecibar, M., Coosemans, T., Messagie, M., Jemal, T., Behabtu, H.A., Van Mierlo, J. (2020). A techno-economic optimization and performance assessment of a 10 kWP photovoltaic grid-connected system. *Sustainability*, 12, 7648.
44. Aneke, M., Wang, M. (2016). Energy storage technologies and real life applications—A state of the art review. *Applied Energy*, 179, 350–377.
45. Ben Elghali, S., Outbib, R., Benbouzid, M. (2019). Selecting and optimal sizing of hybridized energy storage systems for tidal energy integration into power grid. *Journal of Modern Power Systems and Clean Energy*, 7, 113–122.
46. IEC Electrical Energy Storage White Paper (2020). Available online: https://www.iec.ch/whitepaper/energystorage/.
47. Khalid, M. (2019). A review on the selected applications of battery-supercapacitor hybrid energy storage systems for microgrids. *Energies*, 12, 4559.

48. Mallick, K., Das, S., Sengupta, A., Chattaraj, S. (2019). Modern mechanical energy storage systems and technologies. *International Journal of Engineering Research*, 5, 727–730.

49. Kampouris, K.P., Drosou, V., Karytsas, C., Karagiorgas, M. (2020). Energy storage systems review and case study in the residential sector. *IOP Conference Series: Earth and Environmental Science*, 410, 012033.

50. Amiryar, M.E., Pullen, K.R. (2017). A review of flywheel energy storage system technologies and their applications. *Applied Science*, 7, 286.

51. Salkuti, S.R., Jung, C.M. (2018). Comparative analysis of storage techniques for a grid with renewable energy sources. *International Journal of Engineering and Technology*. 7, 970–976.

52. Argyrou, M.C., Christodoulides, P., Wongwises, S.A. (2018). Energy storage for electricity generation and related processes: Technologies appraisal and grid scale applications. *Renewable and Sustainable Energy Reviews*, 94, 804–821.

53. Molina, M.G. (2017). Energy storage and power electronics technologies: A strong combination to empower the transformation to the smart grid. *Proceedings of IEEE*, 105, 2191–2219.

54. Swierczynski, M., Teodorescu, R., Rasmussen, C.N., Rodriguez, P., Vikelgaard, H. (2010). Overview of the energy storage systems for wind power integration enhancement. In *Proceedings of the 2010 IEEE International Symposium on Industrial Electronics*, Bari, Italy; pp. 3749–3756.

55. Suvire, G.O., Mercado, P.E., Ontiveros, L.J. (2010). Comparative analysis of energy storage technologies to compensate wind power short-term fluctuations. In *Proceedings of the 2010 IEEE/PES Transmission and Distribution Conference and Exposition: Latin America, T and D-LA*, São Paulo, Brazil, 8–10 November 2010; pp. 522–528.

56. Asian Development Bank. (2018). *Handbook on Battery Energy Storage System*. Mandaluyong, Phillippines: ADB.

57. Evans, A., Strezov, V., Evans, T.J. (2012). Assessment of utility energy storage options for increased renewable energy penetration. *Sustainable Energy Reviews*, 16, 4141–4147.

58. Stroe, D.I., Stan, A.I., Diosi, R., Teodorescu, R., Andreasen, S.J. (2020). Short term energy storage for grid support in wind power applications. In *Proceedings of the 2012 13th International Conference on Optimization of Electrical and Electronic Equipment (OPTIM)*, Brasov, Romania, 24–26 2012 May; pp. 1012–1021.

59. Mongird, K., Viswanathan, V.V., Balducci, P.J., Alam, M.J.E., Fotedar, V., Koritarov, V.S., Hadjerioua, B. (2019). Energy storage technology and cost characterization report.

60. Yekini Suberu, M., Wazir Mustafa, M., Bashir, N. (2014). Energy storage systems for renewable energy power sector integration and mitigation of intermittency. *Renewable and Sustainable Energy Reviews*, 35, 499–514.

61. Sufyan, M., Rahim, N.A., Aman, M.M., Tan, C.K., Raihan, S.R.S. (2019). Sizing and applications of battery energy storage technologies in smart grid system: A review. *Journal of Renewable and Sustainable Energy*, 11, 014105.

62. Pomper, D.E. (2020). Electricity storage: Technologies and applications. Available online: http://nrri.org/?wpdmdl=700.

63. Chowdhury, M.M., Haque, M., Aktarujjaman, M., Negnevitsky, M., Gargoom, A. (2011). Grid integration impacts and energy storage systems for wind energy applications—A review. In *Proceedings of the 2011 IEEE Power and Energy Society General Meeting*, Detroit, MI, USA; pp. 1–8.

64. Farhadi, M.; Mohammed, O. (2016). Energy storage technologies for high-power applications. *IEEE Transactions on Industry Applications*, 52, 1953–1961.

5 Clean and Green Energy Fundamentals

J. Dhanaselvam and V. Rukkumani
Sri Krishna College of Technology

T. Chinnadurai
Reva University

K. Saravanakumar
Sri Krishna College of Technology

M. Karthigai Pandian
GITAM School of Technology

CONTENTS

DOI: 10.1201/9781003242277-5

5.1   INTRODUCTION

Fossil fuels have been one of the most promising energy resources for more than a century in many of the countries in the world [1]. They help us in generating energy for both domestic and commercial applications. According to BP Statistical Review of World Energy 2020, the total energy production from fossil fuels is 84%, oil is 32%, coal is 27%, and natural gas is 24%. Even though they power the world, it is time to list out some disadvantages: (i) They pollute the environment; (ii) they are non-renewable; and (iii) they require complex process to generate power.

Burning of fossil fuels lead to generation of huge amount of smoke and toxic particles that are harmful to the environment and also to the human. Some toxic gases such as sulfur dioxide cause acid rain, which affects the soil. They affect the human's respiratory system and are the main reason for cancer diseases. The release of carbon dioxide leads to global warming. Another important drawback of these energy sources is they are limited in nature and expensive. Since their generation is a complex process, and they are generated in +a controlled environment, most of the time there is a possibility of accident. So, it is time to move toward low-carbon, clean, and green energy to save the earth.

Electricity can also be generated from natural resources such as sunlight, wind, and water. This type of energy is called green energy. Unlike fossil fuels, they do not harm the environment. On the other hand, clean energy is considered as zero emission sources and they do not emit greenhouse gases and pollute the atmosphere. They are important because (i) the amount of greenhouse gas emission is very minimum than fossil fuels and (ii) they have economic benefits in terms of employment opportunities. As per the statistics, approximately 18 million jobs were offered during the year 2018–2019 [1,2]. The natural resources are commonly considered as green and clean energy sources, and they are used as the most promising potential resources for many applications including heating of the buildings, industrial processes such as chemical and cement plants. [2].

In this chapter, the recent trends and the fundamental concepts of solar energy, hydroelectric energy, wind power, and biomass energy are explained in detail. And also, the associated concepts such as advancement in photovoltaic system, environmental and social impacts, and various applications are also discussed [3].

This chapter starts with the energy from the sun, which is generally regarded as the energy driving the earth. Only one percentage of solar energy is being utilized, and it is more than enough to power the world. Usually, this energy is converted into either heat energy or electrical energy. Based on the usage of moving parts or mechanical

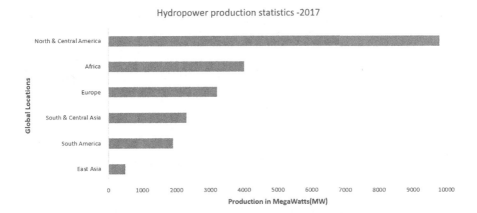

FIGURE 5.1   Hydropower global statistics (see [5]).

devices, they are classified into active and passive solar energy. Photovoltaic systems are the heart of solar energy harvesting process. When the PV systems are technically compatible, they can meet the energy utilization and demand [4].

Another technology in the green energy is hydroelectric energy, where the electricity is generated from the flowing water and turbine systems. Only 25% of the dams in the world are producing electricity. At the same time, others are used for some commercial applications. This is one of the largest power generating sectors. As per the statistics, more than 3,500 hydropower plants are being built and it is planned to double the count by 2030. The following chart tells that in 2017, the production was increased [5] (Figure 5.1).

The next resource is the wind energy. The wind is the result of the sun's uneven heat distribution. In areas with constant and strong winds, wind turbines can be used to generate electricity. Wind energy does not produce air pollution, can be virtually limitless, and is relatively inexpensive to produce. The main disadvantage is wind is not available at all the time and windmills are noisy. The recent advancements are grid-connected wind turbines, which can be used for all the applications from running small, simple machines to powering large, highly sophisticated devices. Moreover, optimization of turbine design, plant design, and specifications is to be concentrated for further advantage.

5.2   SOLAR ENERGY

Solar energy is the most promising energy resource among all because it is inexhaustible in nature. Only one percentage of solar energy is being utilized, and it is more than enough to power the world. Usually, this energy is converted into either heat energy or electrical energy [6]. The sun has a surface temperature of 5,600°C, and the core temperature is 15,000,000°C. Out of all the renewable energy sources, the energy from the sun is considered as the mass potential energy [7]. The principle of generating electricity from the sun is the photovoltaic effect. Silicon-based solar cells are the mostly used for converting solar energy into electricity when more photons

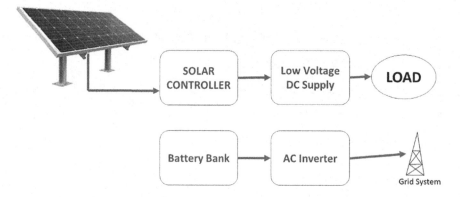

FIGURE 5.2   Stand-alone solar electric system.

are observed by the material. The other materials used for making solar cells are Si, GaAs, and CdS [8]. Recently, some advanced solar cells such as dye-sensitized solar cells, organic solar cells, polymer solar cells, and quantum dot solar cells have been used to tap more energy efficiently [9]. A simplified stand-alone photovoltaic system is shown in Figure 5.2.

The solar energy system is classified into active and passive systems. The active system requires external energy to drive the pump and mechanical devices, whereas the passive system does not require any input energy to the parts and also it is a static system [9].

5.2.1   ACTIVE AND PASSIVE SYSTEMS

The active solar energy system is based on a heat transfer medium and moving parts. Air or liquids are used as a heat transfer medium that collects solar thermal energy. The active system can compensate for other sources or minimize electricity bills for approximately 7 years. According to International Energy Agency (IEA), 45% of the energy demand can be satisfied by solar energy. Most of the industrial applications such as textile, chemical, and food industries are using solar energy as the main resource [10]. Photovoltaic technology and solar thermal technology are the major categories of the active system. In the passive solar system, the energy is stored in the form of distributed heat and it involves collection of energy, storage, and conversion of stored energy. The applications are not limited, and they are extended to space and water heating, temperature swing reduction, and solar cookers [10] (Figure 5.3).

5.2.2   SOLAR PHOTOVOLTAIC TECHNOLOGY (SPVT)

Solar photovoltaic technology is a fast growing clean energy technology. Wafer crystalline silicon and thin-film cells are the major parts of SPVT [11,12]. The system must be economically feasible, and it should have positive impacts on the environment.

The balance of system components is a crucial part to have better life of PV modules. Figure 5.4 shows the fundamental block diagram of a grid-connected solar PV

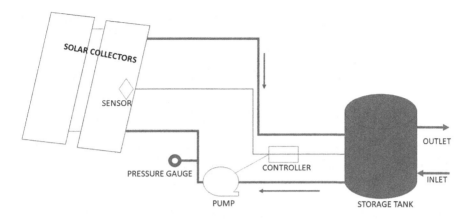

FIGURE 5.3 Active photovoltaic system.

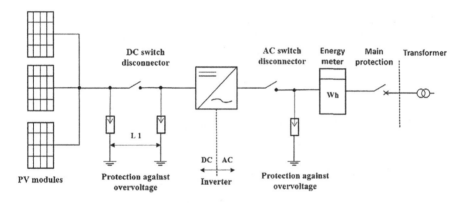

FIGURE 5.4 Grid-connected SPVS.

system with battery arrangement. The expected power delivery is measured in terms of peak kilowatt power. In general, the SPV system varies based on domestic and commercial applications from few kWp to some tens of GWp. The PV system includes PV modules, mounting structure, converters, and batteries. A large number of PV modules are combined together and also integrated into buildings to generate electricity [13].

According to statistics from Renewable Energy World, the solar capacity has increased by 438 MW in Q3, 2020. When compared to Q2, 2020, it is 114% increase [14–15] (Figure 5.5).

5.3 HYDROELECTRIC ENERGY

The hydroelectric power is one of the most widely used clean and green energies in the early 21st century. The potential energy of water falling from some height is converted into electrical energy by using turbines and generators [16] (Figure 5.6).

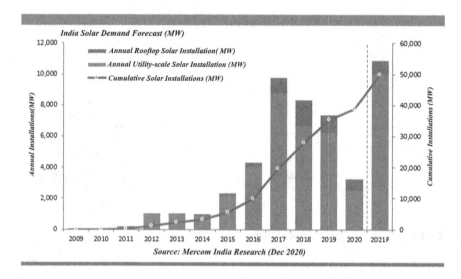

FIGURE 5.5 Indian solar demand forecast statistics.

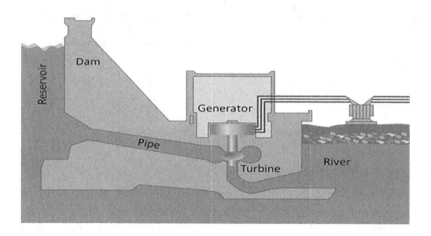

FIGURE 5.6 Structural diagram of hydropower plant.

Figure 5.7 shows the block diagram of electricity generation using hydropower plants. The huge pipe connections are made to connect the top and bottom elevation so that a large amount of water flows from the reservoir and makes the turbine rotate at a high speed. The mechanical energy produced by the turbine is converted into electromechanical energy by the generator. The entire plant arrangement shown in the figure is called a powerhouse [17,18].

The important advantages of hydroelectric energy are as follows: (i) it is non-polluting and clean; (ii) fuel requirement is less; (iii) greenhouse gas emission is very low; (iv) flood can be controlled; and (iv) water can be used for irrigation. It also has some disadvantages:

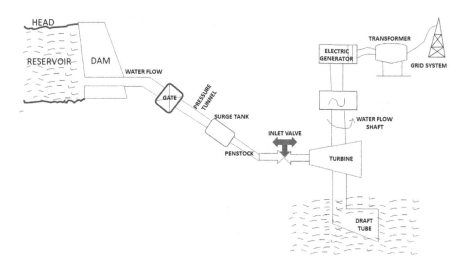

FIGURE 5.7 Flow diagram of HPP.

(i) initial installation cost is very high; (ii) the time period for the construction is large; (iii) population displacement may happen; (iv) during drought season, power output will be very less or there will be no power generation; (v) dam failure will lead to huge problems; and (vi) it is quit impossible to construct dam everywhere [19–20].

5.3.1 HYDROPOWER IN INDIA

India generates clean renewable energy using hydropower plants, and also it stands among the top five in the world. The total capacity is about 50 GW. The countries that participate in hydroelectricity production are China, Brazil, and Canada globally. India has around 197 hydropower plants, and 90% of the plants are operated by public sectors such as National Hydroelectric Power Corporation and North Eastern Electric Power Corporation Limited. The Tehri hydropower dam is the largest plant in India and generates 2,400 MW capacity of power. Also, it is the eighth tallest plant globally and the second in Asia. Next to Tehri hydropower dam, Koyna Hydroelectric Project is the second largest plant in India. The total capacity of the plant is 1,960 MW. The Srisailam Dam is the next largest plant, and it is owned by the Andhra Pradesh Government. The total capacity is about 1,670 MW. The fourth largest hydropower plant is Nathpa Jhakri Dam, which is located at Himachal Pradesh, and the total capacity of the plant is 1,530 MW. And, the fifth largest plant is Sardar Sarovar Dam and is operated by Sardar Sarovar Narmada Nigam. It is located in Gujarat. The total capacity of the plant is 1,450 MW [21] (Figure 5.8).

5.4 BIOMASS ENERGY

The electricity generated by burning once-living organisms is called biomass energy. Usually, plants and their associated wastes are considered as biomass materials. This

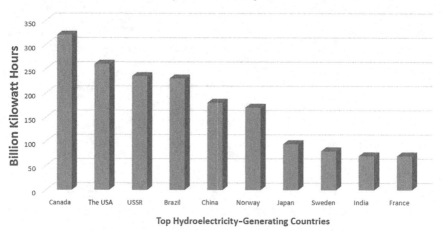

FIGURE 5.8   Global hydroelectricity demand (see [22]).

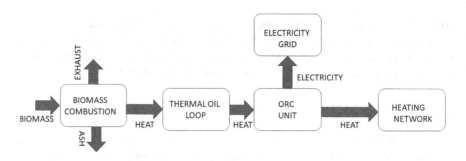

FIGURE 5.9   Biomass energy production process.

energy is also coming under the category of clean and green energy. Biomass energy sources are one of the most promising and heavily subsided energy sources. Among all other energy resources, they have real potential for improving energy security by having very good control over carbon emission into the atmosphere. At the same time, the increased exploitation of biomass materials for energy threatens the environment, water reservoirs, agriculture land, etc. And also, deforestation will be more because of deep exploitation. In the present scenario, it is the need of the hour to discuss the pros and cons of bioenergy as the investment is increasing [23–24]. Figure 5.9 shows the fundamental block diagram of biomass energy production process.

The energy conversion in a biomass plant takes place as follows. They are direct combustion, thermochemical conversion, chemical conversion, and biological conversion, of which direct combustion is the most commonly used process. Some other processes such as gasification and pyrolysis involve thermal decomposition. In this type, a closed and highly pressurized chamber is used to decompose the biomass feedstocks [25].

Sometimes, plant oil and fats of animals are used to produce biodiesel by the process called chemical conversion process. Lastly, the biomass materials are converted into ethanol and natural gas by the process called biological conversion [26].

5.4.1 Biomass and Environment

Carbon is the main element that regulates the temperature of the earth and is the main source of fuels. When carbon atoms travel from atmosphere to earth and earth to atmosphere through all the layers of the earth, it is called carbon cycle [26]. The carbon cycle is described in Figure 5.10.

Carbon helps the environment in many forms. It regulates the amount of sunlight that enters the earth. Soils absorb carbon so that decomposition will take place. Under right pressure and temperature, the decomposing organism can convert carbon into coal and petroleum. The plant and other living organisms take decades to reabsorb the carbon in the carbon cycle. So, the biomass materials are necessary for maintaining a healthy environment [27].

5.4.2 Sources of Biomass Energy

The future energy technology is based on four factors such as conversion technology for the new plant, productive capacity, alternatives to land and water, and offsite implications of biomass energy technologies for invasive species. Before industrial revolution, these types of energy sources were the dominant in the energy field [28].

The source is referred to as heat which is produced from any kind of biological materials. Other resources are water habitats and land areas. The products are categorized into basic firewood to high ethanol or methane. Researchers are working on various types of alternate biomass energy materials, of which engineered

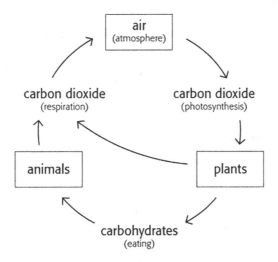

FIGURE 5.10 Carbon cycle.

microorganisms and photosynthetic cells are considered as future biomass energy materials. It is noted that 7% of the world energy production was from biomass during 2000 [29]. The primary applications of biomass are cooking and heating in the developing nations of the world. Another important application is industrial heating, especially in paper industries. Also, biomass energy researches focus on the production of liquid fuels such as ethanol and biodiesel. To meet the requirements of biomass energy, the production must be increased. The net primary production (NPP) of biomass materials determine the quantity of carbon. According to the global statistics, the net quantity of carbon is larger than the industrial consumption. It is shown in Figure 5.10 (Carbon cycle). As per the data, the total carbon emission from the biomass was about 7.7 billion tons in 2005 [30].

Extensively, the biomass as an alternative energy source is used in various sectors such as agriculture, municipalities, micro- and smart grid applications, educational institutions, and hospitals. These energy sources have many potential advantages as follows:

  i. Widely available resource
  ii. Less expensive
  iii. Ability to neutralize carbon
  iv. Reduction in the dependency on fossil fuels
  v. Production of less garbage.

At the same time, they have some limitations that should be acceptable and research is going on to overcome the same.

  i. Not as efficient as fossil fuels
  ii. Deforestation
  iii. Require lot of space
  iv. Not entirely clean.

Even though a lot of research studies are being conducted and a lot of primary applications depend on biomass, still the future prediction is a bit uncertain. They require large amount of space. And also, compared to other green energy sources, they must be economically competitive. So, thinking from this perspective, it is the need of hour to make the future of biomass energy [31,32].

5.5 WIND ENERGY

A new era is ushering in terms of geographic diversity due to the massive changes happening in the geography of renewable energy. To quote an example, only a handful of countries had the privilege of wind power in the 1990s, but now the scenario has changed as almost 82 countries have plunged into power generation using windmills. Wind energy has turned out to be one of the prominent and dependable technologies in terms of renewable energy sources [33]. In 2010, the total capacity of wind turbines installed were 175 GW, which was as 159 GW in 2009.

Wind energy contributes to reducing air pollution caused by greenhouse gases, which results in reduced smog and will not lead to any respiratory issues in human beings. Most of the countries in Europe such as Germany, Netherlands, and Spain have turned to the renewable wind energy for producing electricity [34]. The United States and India are also gradually moving toward implementing a huge number of windmills in their lands. The importance of wind energy lies in its eco-friendly nature [35]. It has been very well documented that the human beings are mainly to be blamed for the considerable discharge of carbon dioxide into the atmosphere. It is observed that the carbon dioxide emission due to human activities reached approximately 2.6 billion tons in the middle of 2002 and it is expected to reach about 4.2 billion in 2030 [36–37]. To overcome this, wind energy can be used as one of the potential alternative energy resources among other conventional sources because it powers human's socioeconomic and industrial activities [38–40]. For optimizing the efficiency of windmills, we need to understand the variation in the net wind power output of the mills in a specific period of time and also the climatic variations. These variations are generally categorized into two types:

1. Short-term variability
2. Long-term variability

5.5.1 SHORT-TERM VARIABILITY

Data analyses performed at operating wind farms and on various atmospheric parameters at specific wind plants help us identify the differential changes in overall wind power output in a given period of time. It could be clearly understood that the short-term variability in power generation could be properly handled by employing efficient power system techniques. In general, the variations in wind power and associated parameters are considered as the relation between the level and variation in power demand. The basic requirement is that the system operators have to continuously monitor the net output of large groups of wind farms.

5.5.1.1 Variations within a Minute

Rapid variations continuously occurring in the output wind power aggregated from a windmill are generally attributed to the turbulence or transient events occurring in its surroundings. These variations (seconds to minute) are very small and hence do not have a great impact on the performance of the windmill.

5.5.1.2 Variations within an Hour

The variations occurring in the aggregated wind power output within an hour are complicated and expected to have a larger impact on the system. They are very closely related to the fluctuations in power demand. The main reason for the local variations is the geographical diversity, which are in the range of ±5% of the total installed capacity. But the most crucial variations are caused by the passage of storm fronts, as the wind turbines are forced to shut down after reaching the cutoff wind speed, which is referred to as the storm limit. In such a scenario, the output power from the windmill swiftly falls to zero, causing a considerable variation in the wind power capacity within the hour. The steep gradient is to be reduced by predicting the passage of storm front, and also, technical solutions are provided based on various wind turbine control mechanisms.

5.5.1.3 Variations from Hour to Hour

These variations are caused by the difference between the forecast predictions and actual wind energy production in a given geographical location that experiences an uncertain weather condition. This in addition will have a huge impact on the power system schedules. Even though these variations do not pose a significant problem, the accuracy in predicting these variations plays a significant role in managing the power capacity schedules of the windmill. A lot of research work is being carried out in identifying the errors in demand forecasts, and solutions are also being worked out to overcome the adversaries.

5.5.2 LONG-TERM VARIABILITY

The long-term variations are considered highly relevant to the wind power system integration, which are completely based on the seasonal climatic effects. These variations are found to have a huge impact on the strategic power system planning.

5.5.2.1 Monthly and Seasonal Variations

In electricity forward contracts, the price per unit power is hugely influenced by the wind power volume, and hence, the electricity traders consider these variations a huge impact factor in their business. Power system planning is another area influenced by these variations.

5.5.2.2 Inter-Annual Variations

These variations are closely related to power system planning on a long-term basis and are not widely considered in daily power operations. The inter-annual variations of the wind power are less than the variations in water inflow in hydropower plants.

5.5.2.3 Characteristics of Wind

Some factors of the atmosphere affect the wind and wind directions. Geography is the main factor, and also, the factors such as time, surface height, seasonal conditions, and land forms are also the reasons for the wind variations. Characterizing the various parameters associated with wind energy can help us in optimizing the design of a wind turbine, in selecting appropriate sites, and in identifying the exact measurement techniques.

5.5.2.4 Wind Speed

Wind speed plays a critical role in the power generation using windmills. Sine waves are generally used to describe the diurnal variations of the average wind speeds over specific geographical locations. As an example, we could see a wavy pattern caused by the diurnal variations of hourly wind speed values that have been recorded at Dhahran, Saudi Arabia [41]. The basic research on wind speed concludes that the wind speed is found to be high during daytime, which describes that the wind speed in daytime is directly proportional to the strength of the sunlight. George et al. [42,43] carried out a research at Lubbock, TX, and identified that the speed of wind shows a curvilinear pattern during daylight hours and was almost constant during night times.

5.5.2.5 Weibull Distribution

The Weibull distribution function is generally used to explain the variation in wind speeds at a given location [44]. The flow of wind speed in a particular geographic location is used to describe the probability of mean speed over a period of time. The probability density function of a Weibull random variable $\bar{u}$ is as follows:

$$f\left(\bar{u},k,\lambda\right) = \begin{cases} \dfrac{k}{\lambda}\left(\dfrac{\bar{u}}{\lambda}\right)^{k-1} \exp\left(-\left(\dfrac{\bar{u}}{\lambda}\right)^{k}\right) & \bar{u} \geq 0 \\ 0 & \bar{u} < 0 \end{cases}$$

From the above expression, it is clearly understood that the mean wind speed and distribution width depend on the scaling factor (λ) and shape factor (k), respectively. Performing statistical analysis on the data collected from the location helps us in determining these two parameters [45,46].

5.5.2.6 Wind Turbulence

Fluctuations occurring in the speed of wind during short time periods are called wind turbulence, especially in the horizontal velocity component. The wind speed $u(t)$ is a dependent variable that depends upon the mean wind speed $\bar{u}$ and the instantaneous speed fluctuation $u'(t)$:

$$u(t) = u + u'(t)$$

The power output fluctuations are mainly caused by wind turbulence, which also reduces the lifetime of the turbine and leads to turbine failure in some cases.

5.5.2.7 Wind Gust

A wind gust is caused because of a sudden drastic change in wind speed with respect to a very small time interval [47]. At the same time, when the wind shear increases, the turbulent gust also increases suddenly.

5.5.2.8 Wind Direction

The direction of wind plays a crucial role in the installation of wind turbine because the amount of energy produced is directly proportional to the speed of the turbine. Some basic analysis can be performed using the wind rose diagram, which helps in understanding the wind data related to wind directions in a specific place over a particular period of time. Figure 5.11 displays the relative frequency of wind directions in 8 or 16 principal directions.

5.5.3 CHALLENGES IN WIND POWER GENERATION

Even though there are a number of advantages offered by wind power generation compared to the conventional methods, it has its own problems to be solved, which could further improve the power generation in windmills.

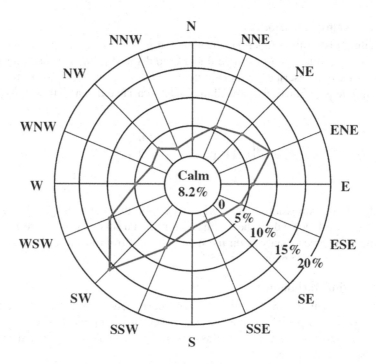

FIGURE 5.11   Wind rose diagram showing wind directions.

5.5.3.1   Impact on the Environmental Conditions

The huge size of the wind turbines employed in modern-day farms and their impact on ecosystem are important issues that have to be addressed. If the site selection has not been properly carried out, these windmills could become an obstruction in the path of migrating birds that could badly be hurt or even be killed. The National Academy of Sciences in the United States reports that an average of 40,000 birds are getting killed due to the wind turbines installed at improper locations.

5.5.3.2   Wind Turbine Noise

In wind turbines, the rotating blades contribute to aerodynamic noise and the gearboxes are supposed to generate mechanical vibration noise, which are the major constituents of wind turbine noises that prove to be an annoying factor for the nearby residents of the area.

5.5.3.3   Integration of Wind Power into Grid

As the contribution of wind energy increases in the global power market, a huge amount of wind power is getting integrated into existing grids. Hence, a situation arises where the current capability of today's electrical grids will not be able to meet the demand posed by the expected growth in wind power generation. To prepare for this situation in advance, the influence of intermittent wind power on the grid stability and system security must be properly addressed.

5.5.4 WIND ENERGY STORAGE

Wind energy could be converted into various other forms of energy and stored; some of them are as follows:

- Electrochemical energy in batteries and supercapacitors
- Magnetic energy in superconducting magnetic energy storage (SMES)
- Kinetic energy in rotating flywheels
- Potential energy in pumped water at higher altitudes
- Mechanical energy in compressed air in vast geologic vaults
- Hydrogen energy by decomposing water.

5.5.5 OFFSHORE WIND TURBINES

One of the recent advancements in wind technology is the offshore wind turbines, which have garnered a huge attraction due to its improved stability. Wind speeds measured across the offshore sea levels are found to be nearly 20% more than the ones measures on the land levels, owing to the low resistance in the environment. Visual and noise pollution is a greater concern in terms of normal windmills, and this is highly reduced with the use of offshore windmills. Transportation of wind structures becomes far easier, and this further contributes to boosting the local economies. Hence, we can conclude that offshore wind turbines may hold a greater advantage in terms of capacity and performance compared to its land-based counterpart.

5.6 CONCLUSIONS

Energy market has been developed in all disciplines. Domestic applications such as home heating and electricity production have a vital role in the present energy market. It is easy at the same time; with minimum investments, people can afford their need. Not only domestic applications, but also commercial and high-level industrial applications demand harmless energy production and eco-friendly energy systems. It is possible only with clean and green energy resources. This chapter is all about the fundamentals and advanced concepts of four major resources such as (i) solar energy and associated active and passive systems, (ii) hydroelectricity fundamentals and global energy market, (iii) biomass energy production and environment consideration, and (iv) wind energy production. Wind integration and its effect on the environment have been discussed. And various merits and demerits have also been taken into consideration.

REFERENCES

1. Wagner, L., I. Ross, J. Foster, and B. Hankamer. "Trading off global fuel supply, CO_2 emissions and sustainable development." *PLoS One* 11, no. 3 (2016): e0149406.
2. Bimanatya, T. E. "Fossil fuels consumption, carbon emissions, and economic growth in Indonesia." (2018).
3. Mouraviev, N., and A. Koulouri. "Clean energy and governance challenges." *International Journal of Technology Intelligence and Planning* 12, no. 1 (2018): 1–5.

 4. Grandell, L., et al. "Role of critical metals in the future markets of clean energy technologies." *Renewable Energy* 95 (2016): 53–62.
 5. Uria-Martinez, R., et al. 2017 Hydropower market report. Oak Ridge National Laboratory (ORNL), Oak Ridge, TN (United States). Oak Ridge National Laboratory Hydropower (ORNLHYDRO), 2018.
 6. Gong, J., C. Li, and M. R. Wasielewski. "Advances in solar energy conversion." *Chemical Society Reviews* 48, no. 7 (2019): 1862–1864.
 7. Boxwell, M. *Solar Electricity Handbook: A Simple, Practical Guide to Solar Energy-Designing and Installing Photovoltaic Solar Electric Systems.* Greenstream Publishing, United Kingdom, 2010.
 8. Hovel, H. J. "Solar cells." *NASA STI/Recon Technical Report A* 76 (1975): 20650.
 9. Fara, L., ed. *Advanced Solar Cell Materials, Technology, Modeling, and Simulation.* IGI Global, Hershey, Pennsylvania, 2012.
10. Mekhilef, S., R. Saidur, and A. Safari. "A review on solar energy use in industries." *Renewable and Sustainable Energy Reviews* 15, no. 4 (2011): 1777–1790.
11. Kabir, E., P. Kumar, S. Kumar, A. A. Adelodun, and K.-H. Kim. "Solar energy: Potential and future prospects." *Renewable and Sustainable Energy Reviews* 82 (2018): 894–900.
12. Besarati, S. M., R. V. Padilla, D. Yogi Goswami, and E. Stefanakos. "The potential of harnessing solar radiation in Iran: Generating solar maps and viability study of PV power plants." *Renewable Energy* 53 (2013): 193–199.
13. Mohanty, P., T. Muneer, E. J. Gago, and Y. Kotak. "Solar radiation fundamentals and PV system components." In *Solar Photovoltaic System Applications*, pp. 7–47. Springer, Cham, 2016.
14. Parida, B., S. Iniyan, and R. Goic. "A review of solar photovoltaic technologies." *Renewable and Sustainable Energy Reviews* 15, no. 3 (2011): 1625–1636.
15. www.renewableenergyworld.com.
16. Sailer, E., et al. "Fundamentals of the coupled thermo-hydro-mechanical behaviour of thermo-active retaining walls." *Computers and Geotechnics* 109 (2019): 189–203.
17. Dincer, I. *Comprehensive Energy Systems.* Elsevier, Amsterdam, Netherlands, 2018.
18. Bahaj, A. S., and L. E. Myers. "Fundamentals applicable to the utilisation of marine current turbines for energy production." *Renewable Energy* 28, no. 14 (2003): 2205–2211.
19. Popp, J., et al. "The effect of bioenergy expansion: Food, energy, and environment." *Renewable and Sustainable Energy Reviews* 32 (2014): 559–578.
20. Baker, N., and K. Steemers. *Energy and Environment in Architecture: A Technical Design Guide.* Taylor & Francis, Boca Raton, FL, 2003.
21. Sharma, N. K., P. K. Tiwari, and Y. R. Sood. "A comprehensive analysis of strategies, policies and development of hydropower in India: Special emphasis on small hydro power." *Renewable and Sustainable Energy Reviews* 18 (2013): 460–470.
22. Bartle, A. "Hydropower potential and development activities." *Energy Policy* 30, no. 14 (2002): 1231–1239.
23. Agbor, V. B., et al. "Biomass pretreatment: Fundamentals toward application." *Biotechnology Advances* 29, no. 6 (2011): 675–685.
24. Van de Velden, M., et al. "Fundamentals, kinetics and endothermicity of the biomass pyrolysis reaction." Renewable energy 35, no. 1 (2010): 232–242.
25. Overend, R. P., T. Milne, and L. Mudge, eds. *Fundamentals of Thermochemical Biomass Conversion.* Springer Science & Business Media, Berlin/Heidelberg, Germany, 2012.
26. Ravindranath, N. H., and D. O. Hall. *Biomass, Energy and Environment: A Developing Country Perspective from India.* Oxford University Press, Oxford, England, 1995.
27. Lynd, L. R. "Overview and evaluation of fuel ethanol from cellulosic biomass: Technology, economics, the environment, and policy." *Annual Review of Energy and the Environment* 21, no. 1 (1996): 403–465.

28. Demirbas, A. "Importance of biomass energy sources for Turkey." *Energy Policy* 36, no. 2 (2008): 834–842.
29. Balat, M., and G. Ayar. "Biomass energy in the world, use of biomass and potential trends." *Energy Sources* 27, no. 10 (2005): 931–940.
30. Chuah, T. G., et al. "Biomass as the renewable energy sources in Malaysia: An overview." *International Journal of Green Energy* 3, no. 3 (2006): 323–346.
31. Field, C. B., J. Elliott Campbell, and D. B. Lobell. "Biomass energy: The scale of the potential resource." *Trends in Ecology & Evolution* 23, no. 2 (2008): 65–72.
32. Wartha, C. "Advantages and disadvantages of biomass fuels on a fundamental combustion basis in fluidized beds." *Fuel and Energy Abstracts* 6, no. 38 (1997): 409.
33. Kose, R., M. A. Ozgur, O. Erbas, and A. Tugcu. "The analysis of wind data and wind energy potential in Kutahya, Turkey." *Renewable and Sustainable Energy Reviews* 8, no. 3 (2004): 277–288.
34. Ahmed, A. S. "Wind energy as a potential generation source at Ras Benas, Egypt." *Renewable and Sustainable Energy Reviews* 14, no. 8 (2010): 2167–2173.
35. Güler, Ö. "Wind energy status in electrical energy production of Turkey." *Renewable and Sustainable Energy Reviews* 13, no. 2 (2009): 473–478.
36. Şen, Z., and A. D. Şahin. "Refined wind energy calculation formulation." In: *World Renewable Energy Congress VI (WREC2000)*, 2000. pp. 2328–2331.
37. Müller-Steinhagen, H., M. R. Malayeri, and A. P. Watkinson. "Heat exchanger fouling: Environmental impacts." *Heat Transfer Engineering* 30, no. 10–11 (2009): 773–776.
38. Wai, R. J., W. H. Wang, and C. Y. Lin. "High-performance stand-alone photovoltaic generation system." *IEEE Transactions on Industrial Electronics* 55, no. 1 (2008): 240–250.
39. Omer, A. M. "On the wind energy resources of Sudan." *Renewable and Sustainable Energy Reviews* 12, no. 8 (2008): 2117–2139.
40. Dincer, F. "The analysis on wind energy electricity generation status, potential and policies in the world." *Renewable and Sustainable Energy Reviews* 15, no. 9 (2011): 5135–5142.
41. Siddiqi, A. H., S. Khan, and S. Rehman. "Wind speed simulation using wavelets." *American Journal of Applied Sciences* 2, no. 2 (2005): 557–564.
42. George, J. M., R. E. Peterson, J. A. Lee, and G. R. Wilson. "Modeling wind and relative humidity effects on air quality." In *International Specialty Conference on Aerosols and Atmospheric Optics: Radiative Balance and Visual Air Quality*, Snowbird, Utah, 1994.
43. George, J. M., G. R. Wilson, and R. C. Vining. "Modeling hourly and daily wind and relative humidity." In *International Conference on Air Pollution from Agricultural Operations*, pp. 183–190, Ames, Iowa, 1996.
44. Weibull, W. "A statistical distribution function of wide applicability." *ASME Journal of Applied Mechanics* 18, no. 3 (1951): 239–297.
45. Ulgen, K., and A. Hepbasli. "Determination of Weibull parameters for wind energy analysis of Izmir, Turkey" *International Journal of Energy Research* 26, no. 6 (2002): 495–506.
46. Yilmaz, V., and H. E. Celik. "A statistical approach to estimate the wind speed distribution: The case of Gelibolu region." *Doğuş Üniversitesi Dergisi* 9, no. 1 (2008): 122–132.
47. Barbu, C., and Vyas, P. System and method for loads reduction in a horizontal-axis wind turbine using upwind information, US patent application, 20,090,047,116, 2009.

6 Evaluation of Sustainable Window for Energy Mitigation in an Electrical Grid Building

Amit Kumar Dhir and Pushpendra Kumar Sharma
Lovely Professional University

CONTENTS

6.1 INTRODUCTION

Energy consumption in the buildings is increasing at an unmatched pace and will reach eight times that of 2012 [1]. Buildings consume 54% share of the total global energy and release 23% carbon dioxide into the atmosphere. The energy demand will rise enormously and needs to be slowed down. Residential buildings consume 25% of the total consumption in India [2]. Thus, the energy demand in buildings has increased. The building envelope is responsible for letting the heat and light inside the building; windows are the critical component of the envelope and are responsible for major heat loss and gain [3]. Thus, it is mandatory to study the energy aspect through the glass system that is integrated as windows to resist the heat gain or loss. On the other hand, the changing life style demands bigger windows for wellbeing and providing comfort to the users [4]. Even the productivity analysis shows betterment in productivity in various built environments such as teaching institutions and hospitals with increased window size; nowadays, it is must to have a larger portion of the envelope to be covered by windows [5]. A balance needs to be maintained with changing life style and energy consumption. The impact of windows should be carefully studied in terms of light efficiency and thermal comfort [6]. Many parameters crucial for windows, such as orientation, climate, and optimal and efficient glass as per climatic conditions, must be carefully considered along with energy analysis to make the user confident [7]. Studies have been performed empirically 68.5% and analytically 11.7%, software 19.8% on

DOI: 10.1201/9781003242277-6

the various aspects of windows making it energy efficient, experimentation is not taken to large level [8,9]. Extensive experimentation is required to ensure that when windows are installed at the consumer location, the results do not deviate from the statistics provided in the manual [7]. Thus, experiments concerning windows for buildings are the need of the hour. Many novel configurations have been used and recommended by researchers [10,11]. An energy saving of 15% is possible using active building envelope technology [12]. Various types of glasses are available, which have the capability to cut the heat ingress into the buildings and reduce the cooling loads in the built sector [13]. Double glazing units have shown better results than traditional units in energy savings in various climates [14–16]. A window should be carefully designed according to key parameters and, in particular, climatic conditions [17]. This chapter evaluates the performance of five different windows, and suitable glazing is recommended as per budget and savings to minimize the energy consumption.

6.2   EXPERIMENTAL SETUP

The objective of this study is to find the most cost-effective windows from the five different windows taken for experimentation. The focus is on the energy consumption criteria that the cooling loads be minimized in the building with the help of economical windows wherein the savings should offset the increased initial cost of the window. To achieve the objective, one single room having dimensions 12 feet × 12 feet × 10 feet is considered in the study. The five different types of windows are listed in Table 6.1. The temperature is recorded with the help of a sensor-based instrument, and readings are taken from 12 to 2 PM on June 21, 2021. The prototype is framed out of plywood having a thickness of ½ inches. The windows are replaced one by one in the provision, and the readings of temperature inside and outside the prototype are recorded. The window is placed at 48% window-to-wall ratio (WWR). The calculation of energy is empirically performed with the help of a 3-star air conditioner having 12,000 BTU. Comparative analyses with respect to the cost of the windows were performed, and the results are further verified by expert vendors. Recommendations are provided to the consumers so that the energy balance can be maintained and also to lower the initial cost with the increased number of window units to lower initial cost. The research procedure is shown in Figure 6.1. Data are gathered from the experimental setup and analyzed for various scopes.

TABLE 6.1
Different Types of Windows

W0CG	Clear Glass 6 mm
W1LEG	Low E Glass 6 mm
W2SCG	Solar Control Glass 6 mm
W3STG	ST167 Glass 12 mm
W4DGU1	Clear Glass 6 mm + 12 mm Air + Clear Glass 6 mm (24 mm width)
W5DGU2	Clear Glass 6 mm + 12 mm Argon + Clear Glass 6 mm (24 mm width)

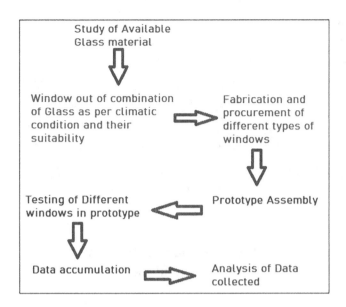

FIGURE 6.1   Research procedure flowchart.

6.3   RESULTS AND DISCUSSION

The experiments were performed in hot and dry climatic conditions. The window that is cost-efficient in cutting the heat is considered and recommended. The different types of windows studied are shown in Table 6.1. The experiments are conducted with the windows facing southward and having a WWR of 48%. The optimum window concerning the payback period should be recommended.

As shown in Figure 6.2, W5DGU2 has the highest performance in maintaining the temperature inside the building with a difference of 8.7°C and the next is W3STG. The reason is the double skin which acts as a barrier to argon gas, and STG type of glass has the ability to resist the heat.

Although the tested windows have the benefit of lowering the temperature inside the building compared to that outside considering the traditionally used windows, the usage is limited in the residential buildings. The reason is the owner's behavior [18] who are not aware of the benefits and types of glass in relation to energy and savings [19]. The challenging part is the initial cost of the windows compared to the traditional windows. Figure 6.3 represents the initial cost of the glass in the windows (the costs of frames and other costs are not taken into account). The windows W3STG, W4DGU1, and W5DGU2 are 3.44–3.7 times costlier than the traditional windows. The initial cost will be offset by the savings with the performance windows.

Figure 6.4 shows the payback period that is due to the savings over the energy consumption. W5DGU2 has the added benefit of argon gas that lowers the payback period to 2.94 years taking 10 months of usage of air conditioner in 1 year.

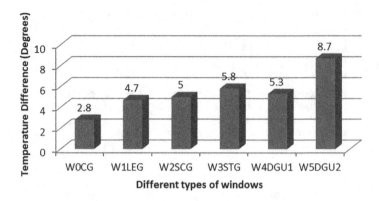

FIGURE 6.2   Difference in temperature for different types of windows.

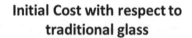

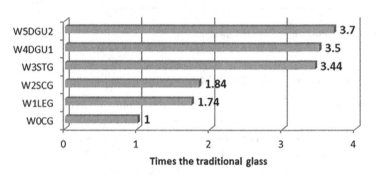

FIGURE 6.3   Initial cost with respect to traditional window.

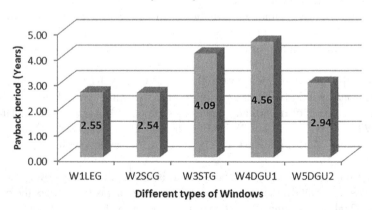

FIGURE 6.4   Payback period of different types of windows.

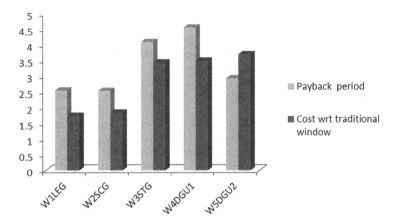

FIGURE 6.5   Comparison of windows.

Figure 6.5 shows the comparison of windows in terms of payback period versus the initial cost of the windows with respect to traditional windows. Seeing the differences between both parameters, W5DGU2 is the suitable window that is recommended to be used in warm and hot climate conditions.

Based on this study, it is suggested that for hot climates, the parameters such as WWR should be 48%, south facing window, the window W5DGU2 which is having clear glass in double skin with argon gas should be used in order to minimize the cooling loads. The recommended solution is also economical because of its payback period.

6.4   CONCLUSIONS

As the outcome indicates, with the usage of inventive windows, the energy consumption can be lowered greatly. In hot climates, the energy consumption increases, which is a challenge for the developing countries to deal with. The results are in agreement with previous studies about the impact of double glazing windows having argon gas. Such a recommendation helps the engineers and designers select suitable windows at early design stages of buildings and also covers the energy breakdown where the initial cost can be offset with the savings.

Out of five different windows studied in comparison with the most widely used and cheap traditional windows, the order of the most suitable windows is W5DGU2, W3STG, and W4DGU1. W5DGU2 has a payback period of 2.94 years and reduces the temperature difference to 8.7°C out of five different performance windows. The cost versus payback period is lesser among all the windows. W3STG is the second viable solution, and its cost is 3.44 times higher, which increases the payback period. The authors will use this window in residential buildings to check for the actual consumption and satisfaction of the owners in future studies. This way the owner feels confident, and the research results can be used by policymakers to develop the nation in a sustainable way. With these proposed windows, the electricity bills can be minimized.

REFERENCES

1. J. Morrissey, "Better data," *Mod. Healthc.*, vol. 27, no. 27, p. 43, 1997, [Online]. Available---http://search.ebscohost.com/login.aspx?direct=true&AuthType=cookie,ip, shib&db=rzh&AN=106102832&site=ehost-live.

2. CSO, "Energy statistics 2019 (twenty sixth issue)," pp. 1–123, 2019, [Online]. Available---http://mospi.nic.in/sites/default/files/publication_reports/EnergyStatistics 2019-finall.pdf.

3. E. Cuce, C. H. Young, and S. B. Riffat, "Performance investigation of heat insulation solar glass for low-carbon buildings," *Energy Convers. Manag.*, vol. 88, no. December 2014, pp. 834–841, 2014, doi:10.1016/j.enconman.2014.09.021.

4. S. Yeom, H. Kim, T. Hong, and M. Lee, "Determining the optimal window size of office buildings considering the workers' task performance and the building's energy consumption," *Build. Environ.*, vol. 177, no. March, p. 106872, 2020, doi:10.1016/j. buildenv.2020.106872.

5. P. Pilechiha, M. Mahdavinejad, F. Pour Rahimian, P. Carnemolla, and S. Seyedzadeh, "Multi-objective optimisation framework for designing office windows---Quality of view, daylight and energy efficiency," *Appl. Energy*, vol. 261, no. July 2019, p. 114356, 2020, doi:10.1016/j.apenergy.2019.114356.

6. R. Alsharif, M. Arashpour, V. Chang, and J. Zhou, "A review of building parameters' roles in conserving energy versus maintaining comfort," *J. Build. Eng.*, vol. 35, no. December 2020, p. 102087, 2021, doi:10.1016/j.jobe.2020.102087.

7. E. Cuce, "Accurate and reliable U-value assessment of argon-filled double glazed windows---A numerical and experimental investigation," *Energy Build.*, vol. 171, pp. 100–106, 2018, doi:10.1016/j.enbuild.2018.04.036.

8. M. Casini, "Active dynamic windows for buildings---A review," *Renew. Energy*, vol. 119. Elsevier Ltd, pp. 923–934, 2018, doi:10.1016/j.renene.2017.12.049.

9. T. J. Dabe and A. R. Dongre, "Analysis of performance of the daylight into critical liveable area of 'type design' dwelling unit on the basis of daylight metrics for hot and dry climate," *Indoor Built Environ.*, vol. 27, no. 1, pp. 129–142, 2018, doi:10.1177/1420326X16669844.

10. R. Tafakkori and A. Fattahi, "Introducing novel configurations for double-glazed windows with lower energy loss," *Sustain. Energy Technol. Assessments*, vol. 43, no. November 2020, p. 100919, 2021, doi:10.1016/j.seta.2020.100919.

11. J. Hirvonen et al., "Emissions and power demand in optimal energy retrofit scenarios of the Finnish building stock by 2050," *Sustain. Cities Soc.*, vol. 70, no. December, 2021, doi:10.1016/j.scs.2021.102896.

12. F. Mokhtari, D. Semmar, M. Chikhi, N. Kasbadji Merzouk, and S. Oukaci, "Investigation of the improvement building envelope impact on energy consumption using energy audit," *MATEC Web Conf.*, vol. 307, p. 01031, 2020, doi:10.1051/matecconf/202030701031.

13. Y. Yaşar and S. M. Kalfa, "The effects of window alternatives on energy efficiency and building economy in high-rise residential buildings in moderate to humid climates," *Energy Convers. Manag.*, vol. 64, no. 4, pp. 170–181, 2012, doi:10.1016/j. enconman.2012.05.023.

14. W. Zhang, L. Lu, and X. Xu, "Thermal and daylighting performance of glass window using a newly developed transparent heat insulated coating," *Energy Procedia*, vol. 158, pp. 1080–1085, 2019, doi:10.1016/j.egypro.2019.01.262.

15. K. Gorantla, S. Shaik, and A. B. T. P. Setty, "Effect of different double glazing window combinations on heat gain in buildings for passive cooling in various climatic regions of India," *Mater. Today Proc.*, vol. 4, no. 2, pp. 1910–1916, 2017, doi:10.1016/j. matpr.2017.02.036.

16. M. A. Fasi and I. M. Budaiwi, "Energy performance of windows in office buildings considering daylight integration and visual comfort in hot climates," *Energy Build.*, vol. 108, pp. 307–316, 2015, doi:10.1016/j.enbuild.2015.09.024.
17. A. Dhir and P. K. Sharma, "A evaluation for role of suitable glazing in saving energy," vol. 6, no. 1, pp. 1030–1035, 2019.
18. S. Jami, N. Forouzandeh, Z. S. Zomorodian, M. Tahsildoost, and M. Khoshbakht, "The effect of occupant behaviors on energy retrofit---A case study of student dormitories in Tehran," *J. Clean. Prod.*, vol. 278, p. 123556, 2021, doi:10.1016/j.jclepro.2020.123556.
19. A. K. Dhir, "Role of suitable glazing in minimizing energy consumption in Jalandhar region," *Int. J. Res. Anal.* vol. 5, no. 4, pp. 1527–1530, 2018.

7 Filter Bank Multicarrier for Smart Grid Systems

Shakti Raj Chopra

Lovely professional University

CONTENTS

7.1 INTRODUCTION

Smart grid is expected to emerge as a potential candidate for the implementation of new generation. It is a multiuser technique to increase the capacity of a multiuser network. Here the gain spread is determined by channel gain between mobile antenna and antennas at the base station. As the number of base station antennas is increased, the processing gain becomes very large. When the number of antennas at a base station approaches infinity, the system processing gain also approaches infinity. The result is that the negative effect of noise and interference due to multiple users are eliminated. So the capacity of the network can theoretically be increased without any bound when antennas at the base station are increased.

An assumption is that OFDM will convert the frequency-selective channel in between MT and base station to a group of flat fading channels. So flat gains that are associated with a group of subchannels of every subcarrier give the spread gain vector that can be used to dispread the data streams.

I introduce the concept of filter bank multicarrier (FBMC) technique to wireless communication system. FBMC does not require cyclic prefix and takes up a number of subcarriers. It depends on the fact that if each subcarrier has a narrow band, approximately flat gain can be achieved with minimum intersymbol interference [1].

The technology that currently dominates the field of communication using broadband multicarrier techniques is the orthogonal frequency division multiplexing (OFDM). Be that as it may, in specific applications, for example, cognitive radios and also uplink of multicarrier system involving multiple users, where a subset of

DOI: 10.1201/9781003242277-7

subcarriers are made available to every client, OFDM may not be as useful as it is desired. In this chapter, we address the inadequacies of OFDM in these and other different applications and demonstrate that FBMC could be a more powerful and effective arrangement. Despite the fact that FBMC strategies have been examined by various specialists, and some even before the creation of OFDM, just as of late has FBMC been truly considered by committees of standard boards. The objective of this chapter is to bring this technology to the consideration for signal processing and communication and carry out more research in this area [2–5].

A lot of research has been carried out, and hence, OFDM has been used for both wired and wireless communications in broadband for past several years. A lot of advantages are offered when using OFDM, particularly:

1. It allows subcarrier signals to be orthogonal, which helps in trivially:
   - generating signals for transmission using inverse fast Fourier transform (iFFT) block;
   - separating the data symbols that were transmitted using a block of FFT at the receiver;
   - equalizing using a scalar gain per given subcarrier; and
   - adopting to channels of multiple input and multiple output type.

2. The available bandwidth is divided into a maximum number of narrow sub-bands. This is possible due to close spacing of the orthogonal subcarriers.
3. To maximize bandwidth efficiency or transmission rate, adaptive modulation techniques can be used over subcarrier bands.
4. Carrier and symbol synchronization becomes easy due to OFDM structure.

7.2 FILTER BANK MULTICARRIER

The filter bank multicarrier (FBMC) transmission system prompts an improved physical layer for ordinary correspondence systems, and it is an empowering innovation for the new ideas and, especially, psychological radio. The goal of this chapter is to give an outline of FBMC, with accentuation on the highlights which affect the corresponding systems. The main essential for perusing this chapter is essential learning in advanced flag preparing, specifically examining hypothesis, fast Fourier transform (FFT), and finite impulse response (FIR) separating.

The introduction starts with the immediate use of FFT to multicarrier correspondences, bringing up the confinements of this short-sighted approach and, especially, the range spillage. At this point, it is demonstrated that the FFT approach can develop a channel bank approach, which is easy to plan and execute. For each piece of information, the time window is stretched past the multicarrier image period and the images covered in the time period. This time covering is at the premise of customary productive single transporter modems where impedance between the images is stayed away from if the channel fulfils the Nyquist foundation. This major rule is prompt material to multicarrier transmission. As to, the channel bank approach is only an augmentation of the coordinate FFT approach and it can be acknowledged with an expanded FFT. An elective plan, requiring less calculations, is the supposed polyphase system (PPN)–FFT strategy, which keeps the measure of the FFT yet include an arrangement of computerized channels.

In opposition to OFDM where orthogonality must be guaranteed for every one of the transporters, FBMC requires orthogonality for the neighboring subchannels as it were. Truth be told, OFDM abuses the given recurrence data transfer capacity with various transporters, while FBMC separates the transmission channel related to this given data transmission into a number of subchannels. Keeping in mind the end goal to completely abuse the channel transfer speed, the adjustment in the subchannels must adjust to the neighbor orthogonality imperative and counterbalance quadrature adequacy adjustment (OQAM) is utilized for that reason. The blend of channel saves money with OQAM regulation prompting the greatest piece rate, without the requirement for a watch time or cyclic prefix as in OFDM.

The impacts of the transmission channel are enumerated at the subchannel level. The subchannel equalizer can adapt to bearer recurrence counterbalance, timing balance, and stage and adequacy mutilations, with the goal that non-concurrent clients can be accommodated. At the point when FBMC is utilized in burst transmission, the length of the burst is reached out to take into consideration the initial and last changes because of the channel motivation reaction. These advances might be abbreviated assuming a few brief recurrence spillage is permitted, for instance at whatever point a recurrence hole is available between neighboring clients. As a multicarrier technique, FBMC can profit by multi-antenna frameworks and MIMO methods can be adopted. Because of OQAM adjustment, adjustments are essential for some MIMO approaches, in the assorted variety setting.

FBMC frameworks are probably going to exist together with OFDM frameworks. Since FBMC is a development of OFDM, some similarity can be normal. Indeed, the instatement stage can be normal to both and effective double mode usage can be figured out.

In the multiuser setting, the subchannels or gatherings of subchannels assigned to the clients are frightfully isolated when a vacant sub-direct is available in the middle. Accordingly, clients should not try to be synchronized before they access the transmission framework. This is an essential office for uplink in base station-ruled systems or for future astute interchanges. In intellectual radio, the FBMC system offers the likelihood to complete the elements of range detection and transmission with a similar gadget, together and all the while. In addition, the clients appreciate an ensured level of protection of spectrum [6–10].

7.3 FAST FOURIER TRANSFORM AS MULTICARRIER MODULATOR

The inverse fast Fourier transform (iFFT) can fill in as a multicarrier modulator, and the fast Fourier transform (FFT) can act as a multicarrier demodulator. A multicarrier transmission framework is acquired, and the transmitter and receiver are shown in Figure 7.1.

It is clear from the assumption that the square of information at the contribution of the iFFT in the transmitter is recouped at the yield of the FFT in the beneficiary, since the FFT and the iFFT are put back to back.

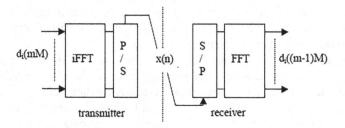

FIGURE 7.1   FFT as a multicarrier modulator.

The point-by-point depiction of the tasks is as follows.

The measure of the iFFT and the FFT is M, and an arrangement of M information tests, $d_i(mM)$ with $0 \leq I \leq M-1$, is bolstered to the iFFT input. For $mM \leq n < (m+1)M$, the iFFT yield is communicated by

$$x(n) = \sum_{i=0}^{M-1} d_i(mM) e^{j2\pi i(n-mM)/M}$$

The arrangement of M tests so acquired is known as a multicarrier image, and m is the image record. For transmission in the channel, a parallel-to-serial (P/S) converter is presented at the yield of the iFFT, and the examples $x(n)$ show up in serial shape. The examining recurrence of the transmitted flag is solidarity, there are M bearers, and the transporter recurrence dispersing is $1/M$. The span of a multicarrier image T is the reverse of the transporter dividing, $T = M$. Note that T is likewise the multicarrier image period, which mirrors the way that progressive multicarrier images don't cover in the time space.

A representation is given in Figure 7.2 for $I = 2$ and $(\cdot)$ 1 2 d $mM = \pm 1$. The transmitted signal $x(n)$ is a sine wave, and the term T contains $I = 2$ periods. Correspondingly, $d_i(mM)$ is transmitted by i times of a sine wave in the span T. By and large, the transmitted flag is an accumulation of sine waves with the end goal that the image term contains a whole number of periods. Truth be told, it is the condition for information recuperation, the supposed orthogonality condition.

At the reception, a serial-to-parallel (S/P) converter is presented at the contribution of the FFT. The information tests are carried out by

$$d_i(mM) = \frac{1}{M} \sum_{n=mM}^{mM+M-1} x(n) e^{-j2\pi i(n-mM)/M}$$

Note likewise in Figure 7.1 that, because of the course of P/S and S/P converters, there is a postponement of one multicarrier image at the FFT yield as for the iFFT input [11].

For the best possible working of the framework, the beneficiary (FFT) must be superbly adjusted in time with the transmitter (iFFT). Presently, within the sight of a channel with multipath proliferation, because of the channel motivation reaction, the multicarrier images cover at the recipient input also and it is not any more

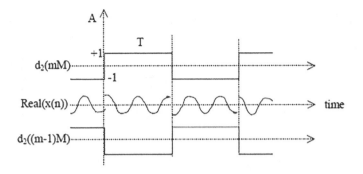

FIGURE 7.2 Data and transmitted signal.

conceivable to demodulate with simply the FFT, on the grounds that intersymbol obstruction has been presented and the orthogonality property of the transporters has been lost [12].

At this point, there are two choices:

1. Broaden the image term by a protect time surpassing the length of the channel motivation reaction and still demodulate with the same FFT. The plan is called OFDM.
2. Keep the planning and the image length as they may be, yet add some preparing to the FFT. The plan is called FBMC, on the grounds that this extra handling and the FFT together constitute a bank of channels.

This chapter is a discussion about this second approach, and as a presentation, it will in the first place be demonstrated that the FFT itself is a channel bank.

7.4 FFT'S FILTERING EFFECT

Give us a chance to expect that the FFT is running at the rate of the serially transmitted examples. Thinking about Figure7.1, the connection between the contribution of the FFT and the yield with list $k=0$ is the accompanying

$$y_0(n) = \frac{1}{M}[x(n-M)+\cdots+x(n-1)]$$

This is the condition of a low-pass straight stage FIR channel with M coefficients equivalent to $1/M$. Regardless of the steady deferral, the recurrence reaction is

$$I(f) = \frac{\sin \pi f M}{M \sin \pi f}$$

It appears in Figure 7.3, where the unit on the recurrence hub is $1/M$.

In similar conditions, the FFT yield with list k is communicated by

$$y_k(n) = x(n-M+i)e^{-j2\pi ki/M}$$

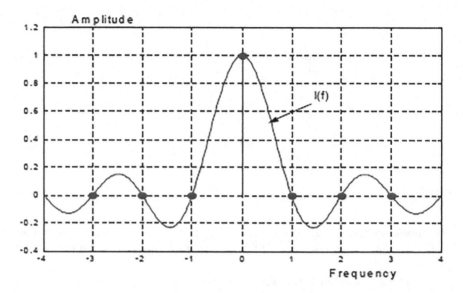

FIGURE 7.3   FFT filter response and coefficient in frequency domain.

Changing variables and replacing i by $M - i$, an alternative expression is

$$y_k(n) = \frac{1}{M}\sum_{i=1}^{M}x(n-i)e^{j2\pi ki/M}$$

The channel coefficients are increased by $e^{j2\pi\,ki/M}$, which compares to a move in recurrence by k/M of the recurrence reaction. At the point when all the FFT yields are viewed as, a bank of M channels is acquired, as appeared in Figure 7.4, in which the unit on the recurrence hub is $1/M$, the sub-transporter separating. The orthogonality condition shows up through the zero intersections: At the frequencies which are number products of $1/M$, just a single channel recurrence reaction is non-zero.

A FIR channel can be characterized by coefficients in the time period or by coefficients in the recurrence space. The two arrangements of coefficients are proportional and related by the discrete Fourier transform (DFT). Coming back to the principal channel in the bank, the DFT of its motivation reaction comprises of a solitary heartbeat, as shown in Figure 7.3. Truth be told, the recurrence coefficients are the examples of the recurrence reaction $I(f)$, which, as per the inspecting hypothesis, is obtained from them through the introduction equation.

In the wording of channel banks, the main channel in the bank, the channel related to the zero recurrence transporter, is known as the model channel, in light of the fact that alternate channels are concluded from it through recurrence shifts. It is clear in Figure 7.4 that $I(f)$ is the recurrence reaction of a model channel with constrained execution, especially out-of-band weakening. Keeping in mind the end goal to lessen the out-of-band swells, it is important to build the quantity of coefficients in the time period and, equally, in the recurrence space. At this point, in the time space, the channel motivation reaction length surpasses the multicarrier image period T. In the

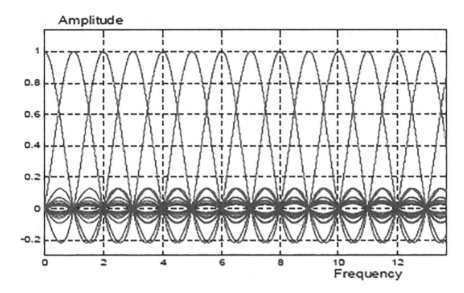

FIGURE 7.4 FFT filter bank.

recurrence space, extra coefficients are embedded between the current coefficients, taking into account a superior control of the channel recurrence reaction.

Model channels are described by the covering factor K, which is the proportion of the channel motivation reaction length Θ to the multicarrier image period T. The factor K is additionally the number of multicarrier images which are covered in the time space. By and large, K is a whole number and, in the recurrence space, it is the quantity of recurrence coefficients which are presented between the FFT channel coefficients.

Presently, the inquiry is the means by which to outline the model channel and transmit information in such a way that no intersymbol impedance happens, disregarding the covering.

7.5 SYSTEM MODEL AND FBMC
FORMULATION IN SMART GRID

Definite data on various sorts of FBMC frameworks can be found in the book by Farhang-Boroujeny [13]. Among the diverse sorts of FBMC framework, we are more intrigued in FBMC frameworks with covering subcarriers in the recurrence space to be SMT (specific amazed multitone) and CMT, as these give the most astounding transfer speed effectiveness. Both CMT and SMT can be embraced for MIMO, prompting a similar execution. In any case, it turns out that inference and clarification of the outcomes with regard to CMT are less demanding to take after. We subsequently restrain our consideration in whatever is left of this section to the advancement of CMT in monstrous MIMO applications. There are two unique usages of FBMC frameworks [14–16], in view of polyphase systems (PPN) and recurrence spreading ideas, separately. In CMT, an arrangement of pulse amplitude regulated

(PAM) baseband information streams are minimal sideband (VSB) adjusted and put at various subcarriers. Also, to permit partition of the information images (free of ISI and ICI), at the collector, the transporter period of the VSB signals is flipped in the vicinity of 0 and $\pi/2$ among nearby subcarriers. The nitty gritty conditions clarifying why this approach works can be found in a 1966 paper by Chang [18] and numerous different distributions; a prescribed reference is Farhang-Boroujeny's 2010 paper [13]. That writer's book [17] gives more subtle elements, including the usage structures and their relevant MATLAB® codes. Demodulation of each subcarrier in CMT is performed in four stages.

1. The receiving signal is down-changed over to baseband utilizing the comparing bearer recurrence to each subcarrier.
2. The demodulated flag is sent through a coordinated channel that concentrates the coveted flag at the baseband. Because of the cover among the contiguous subcarriers, a few residuals from contiguous subcarriers stay after coordinated separation.
3. A complex esteemed single-tap equalizer is used to even out the channel effect.
4. As the genuine piece of the leveled flag contains the coveted PAM image and its fanciful part comprises of a blend of ISI segments and ICI segments from the two contiguous groups, taking the genuine piece of the adjusted flag conveys the coveted information image, free of ISI what's more, ICI.

7.6 SELF-EQUALIZATION PROPERTY OF FBMC IN SMART GRID

In traditional (single-input, single-yield) FBMC frameworks, keeping in mind the end goal to lessen channel leveling to single tap per subcarrier, it is frequently expected that the quantity of subcarriers is extremely extensive, so each subcarrier band might be approximated by a level pickup. This obviously has the unwanted impact of lessening the image rate (per subcarrier), which carries with it:

- the requirement for longer pilot prefaces (equally, a decrease in the data transmission productivity)
- increment of inertness in the channel
- higher affectability to transporter recurrence balance (CFO)
- higher crest-to-normal power proportion (PAPR) because of the substantial number of subcarriers, which builds the dynamic scope of the FBMC flag.

Huge MIMO channels have an intriguing property that enables us to determine the above issues. The MF and MMSE indicators that are utilized to consolidate signals from the get reception apparatuses normal the bends from various channels and, in this way, as the quantity of BS receiving wires increases, result in an about balanced pickup over each subcarrier band. This property of monstrous MIMO channels is called self-evening out.

7.7   CHOICE OF FILTER BANK STRUCTURE

The trans-multiplexer configuration forms the heart of FBMC system. This representation is of direct form. OQAM preprocessing block, synthesis and analysis filter banks, and OQAM post-processing blocks are the main blocks of this structure. As the equalization of channels is dealt with separately, the transmission channel is left out when the TMUX systems are being analyzed and designed.

Only those M-channel filter banks are studied, which can be classified according to the following:

- **Complex Modulation**: Only complex in/quad phase baseband signal can deliver a good spectral efficiency. That is why these kinds of signals are needed for transmission purposes. This makes complex modulated filter banks a favorable choice, meaning subchannel filters are frequency-shifted versions of the given prototype filter.
- **Uniformity**: The same bandwidth is found for all types of subchannel filters as the given prototype filter, and it divides equally whatever the bandwidth for the channel is available.
- **Finite Impulse Response**: The reason why finite impulse response filters are chosen is that they are easy to design and implement.
- **Orthogonality**: Only one prototype filter is required for orthogonal type of filter banks, and so subchannel filters which are linear phased type can be obtained using exponential schemes of modulation. The order of the filter determines the delay of the overall system.
- **Nearly Perfect Reconstruction**: Signals at the output are just the delayed versions of the signals at the input. So distortions due to filter banks can be tolerated as long as they are within specified limits compared to distortion due to transmission channel.

Here the objective is to discuss a method of designing filter banks using optimization techniques. As discussed above, the main objective is to achieve perfect reconstruction of the signals. Also there are some frequency constraints that are needed to be satisfied. The method of approach forms its basis by having the design problem being formulated as an optimization problem. Here a parameter of performance is needed with one part describing the perfect reconstruction of signal and the other part describing magnitude specifications for the receiving filter. A set of linear equations using Z-transform is used for reconstructing perfectly the signal obtained from FIR transmission and receiving filters.

An essential limitation of information transmission is that the channel must fulfill the Nyquist criterion, to maintain a strategic distance from intersymbol impedance. On the off chance that the image period is T_{symb} and the image rate is $f_{symb} = 1/T_{symb}$, the channel recurrence reaction must be symmetrical about the recurrence $f_{symb}/2$. In a similar manner, in FBMC, the model channel for the blend and investigation channel banks must be half-Nyquist, which implies that the square of its recurrence reaction must fulfil the Nyquist paradigm.

In this chapter, we consider uniform channel banks; that is, all the subchannels have a similar data transmission. Proficient uniform channel banks can be executed utilizing different structures using balance to make band-pass subchannels from a solitary low-pass model channel, fundamentally through recurrence moving. There are different proficient multi-rate structures for the required channel banks, counting lapped changes, cross-sectional structures, and the polyphase structure [18–20]. Basically, all these structures comprise of a channel segment, the coefficients of which are controlled by the model channel plan and a change area (e.g., discrete Fourier, sine, or cosine changes) executing the regulation. In mix with the change obstructs, the structures incorporate examining rate change activities, with the end goal that the subchannel signals work at the essential signaling rate, while the orchestrated wideband flag has a considerably higher inspecting rate. In a basically tested channel bank framework, the example rate (checked as far as there are genuine esteemed examples in the conceivable complex (I/Q) signals) of the SFB yield and AFB input is equivalent to the entirety of the test rates of the subchannel signals. In the FBMC application, the utilization of fundamentally examined channel banks would be tricky, since the associating impacts would make it hard to adjust defects of the channel by handling the subchannel motions after the AFB as it were. Consequently, a factor of two oversampling is regularly connected in the subchannel motions in the AFB. In the considered channel bank models, the helpful information images are conveyed alternatively by genuine and nonexistent parts of the complex esteemed subcarrier arrangements. By utilizing the entire complex examples in subchannel preparing in the collector, successfully 2× oversampling is acquired. Toward the finish of the subchannel handling segments, the required genuine/nonexistent parts are chosen to get a fundamentally inspected grouping for identification.

Purported idealize recreation (PR) channel banks actualize the Nyquist basis precisely and furthermore without presenting any cross-talk between subchannels in the consecutive association of SFB and AFB (supposed trans-multiplexer). In remote interchanges, the transmission channel presents definitely some bending to obtain subchannel signals. In this manner, the PR condition isn't fundamental, and it is adequate that the cross-talk between subchannels is sufficiently little to be disregarded in contrast to the leftover obstruction, e.g., because of flawed channel evening out. From the channel bank configuration perspective, this implies the supposed about immaculate remaking Nearly Perfect Reconstruction [NPR] plans are adequate. Since NPR plans are more effective than PR outlines, e.g., in giving higher recurrence selectivity with given model channel length, NPR outlines are the favored decision in PHYDYAS. For NPR channel banks, the polyphase structure is the regular decision, since lapped changes and cross-sectional structure can be utilized just in the PR case [21,22].

Amid the main period of the PHYDYAS venture, the work depends on a chosen channel bank outline known from the writing. This channel bank depends on the polyphase structure and logical equations for ascertaining the channel coefficients for a wide decision of the principle parameters:

- The quantity of subchannels (M) is fundamentally subjective; however, ordinarily, an energy of 2 is utilized as a part of request to have the capacity to utilize iFFT/FFT as proficient calculations for the change squares.

- Overlapping factor (K) can be chosen to be 3 or higher. The essential decision for model channel length is $L = KM$, yet additionally, $L = KM + 1$ or $L = KM - 1$ is an intriguing option.
- The move off parameter of this plan is basically $\alpha = 1$, which implies that the change groups of a subchannel end at the focuses of the neighboring subchannels. This implies as it were quickly adjoining subchannels are essentially collaborating with each other.

7.8   GOAL OF WP3

The goal of WP3 is to characterize the handling required at the beneficiary for balance and demodulation in a FBMC framework, remembering the adaptability necessities for multi-client situations. The primary focuses are for the most part straightforwardly identified with the leveling issue. They concern the plan of the equalizer structure most appropriate to the thought about regulation and condition, the investigation of pilot arrangements and measure of pilots required for estimation and synchronization purposes, and lastly the productive demodulation of OQAM signals. Every one of these focuses are considered and talked about in the present deliverable.

Evening out is a basic piece of the handling at the recipient, which is important to deal with the multipath nature of the transmission channel. One of the enormous favorable circumstances of FBMC is that it permits straightforward per-subcarrier leveling, similarly as in OFDM, without the loss of transmission capacity productivity related to the cyclic prefix. FBMC depends on the selectivity of the channel bank channels to moderate the intercarrier obstruction and guarantees that each subcarrier is sufficiently limited to have a basically level channel inside its data transfer capacity. A few remarks are imperative with respect to the balance issue for FBMC. They are as follows.

- **Number of Subcarriers**: It is important to pick an adequate number of subcarriers, with the goal that each individual subcarrier has a tight transmission capacity and can be adjusted effortlessly, with a set number of taps. Furthermore, a high number of subcarriers give a higher determination as far as range detecting, and an expanded adaptability for recurrence division among clients for example. The drawback is that the many-sided quality increments with the quantity of subcarriers and the affectability to synchronization issue are likewise expanded. The burst truncation impacts are likewise harder to relieve when utilizing a higher number of subcarriers. In view of every one of these contemplations, and so as to upgrade the similarity with the current WiMAX frameworks, it has been worked with 1,024 subcarriers as benchmark. This shows up (get comes about underneath) to be adequate to have an extremely basic single-tap evening out on average channels.
- **OQAM Demodulation**: The specific configuration of the OQAM adjustment, transmitting on the other hand on the genuine and nonexistent part, permits to take full favorable position of the transmission capacity and

transmit at high transmission capacity efficiency. Leveling needs, however, to ensure that no intersymbol is produced between the genuine and nonexistent parts because of the channel recurrence selectivity. Once more, with 1,024 subcarriers on somewhat specific channels, the subsequent intersymbol obstruction is generally low. For more particular channels, equalizers with a few taps should be composed. The most straightforward approach to dealing with the OQAM demodulation, with direct equalizers, is portrayed beneath. Despite the fact that the gotten comes about are now attractive, assist upgrades could be conceivable, utilizing non-straight beneficiaries, to consider the data contained in the corresponding (fanciful or genuine) some portion of each got image. This examination is left for future work.

- **Fragmentary Examining**: The image length is here indicated by T (see framework depiction). As the images are really produced at $T/2$ due to the OQAM adjustment arrangement, it is normal for the equalizer to work at the partial sampling $T/2$. It has the benefit of relating to the inexact data transmission of every individual channel of the channel bank, which implies that no association is produced when playing out the adjustment at the fragmentary inspection. Consequently, all the equalizers considered here work in light of the current circumstances. It is hypothetically conceivable to perform evening out at the image rate $1/T$ when utilizing multi-band equalizers, yet it doesn't bring a particular favorable position, so it won't be considered here.

- **Channel Estimation**: Most equalizer plans depend on the supposition that some channel evaluation is accessible. It is along these lines important to examine precise channel estimation. There are unique strategies utilizing pilot images, for which it is obviously essential both to gauge the channel at the introduction and also to track the progressions of the channel due to the portability in a remote domain.

- **Adaptivity**: While channel estimation in light of pilots is important for a quick instatement, the following can be performed utilizing versatile strategies that don't really require an express gauge of the channel. Visually impaired (or choice coordinated) systems can even further diminish the measure of required pilots and increase the net information rate. As an initial move toward this course, a straightforward LMS calculation is portrayed beneath.

- **Synchronization Issues**: Similarly to the issue of channel estimation, it is obviously important to track both the CFO (transporter recurrence balance) and the image timing amid transmission. The following and pay of the CFO is talked about in Section 7.4. Some portion of the following (image) timing can be performed because of the adjustment of the equalizer. In any case, a solitary tap equalizer can't adjust for any planning mistake, so some coarse following of the planning would be required all things considered. This is contemplated in Section 7.4. An equalizer with a few taps has following capacity, permitting to mitigate the exactness limitations on the following plan, so it may enthusiasm to marginally build the equalizer multifaceted nature for this reason.

Multi-client plans: One of the fundamental focal points of the FBMC framework is that it permits a simple numerous entrance conspire by dispensing diverse arrangements of subcarriers to various clients. A recurrence monitor interim of one subcarrier is adequate to isolate the clients, on account of the selectivity of the channels. Every client can apply per-subcarrier equalizers with all other subcarriers having no adjustment, so the evening out stays extremely straightforward. The CFO and timing remuneration bring a more troublesome issue, especially in uplink. Every client has now an alternate CFO and timing, and it is impractical to make up for every one of them together before the examination channel bank. It is obviously conceivable to compel some synchronization of the diverse clients, with input components, as it is done in OFDM. Be that as it may, it would offer significantly more adaptability to the framework to permit unsynchronized clients to transmit at the same time and this will be examined amid the following semester. The mix of single bearer adjustment and multicarrier tweak in the uplink will likewise be contemplated.

- **Leveling Structures**: Different conceivable evening out structures will be depicted, broken down, and thought about underneath. The acquired execution will likewise be contrasted with OFDM to affirm the normal pickup. Here are the diverse equalizer structures examined.
    - The fundamental evening out is a solitary tap for each sub-transporter equalizer. It depends on channel estimation and will by and large simply upset the channel at the middle recurrence of the comparing subcarrier. It is functioning admirably in particular channels as long as the number of subcarriers is adequate.
        - The primary upgrade is to have an equalizer with numerous taps (per-subcarrier), regularly working at $T/2$ and generally restricted to 3–5 taps. It permits adjusting for more specific channels and for timing mistakes. There are a few conceivable plan criteria:
    - A low-multifaceted nature arrangement depends on recurrence introduction (also called recurrence testing). Its rule is to utilize the channel estimation at the middle frequencies of each subcarrier, and after that interject keeping in mind the end goal to give a rough recurrence reaction inside each subchannel, that can straightforwardly be reversed to give a reasonable equalizer.
    - The MMSE rule can be utilized, in view of channel estimation. This is moderately mind boggling as it requires the calculation of the connection framework and additionally its reversal. A rearrangement will be recommended that marginally diminishes the multifaceted nature related to the calculation of the grid.
    - Finally, it is additionally conceivable to utilize versatile techniques that don't depend straightforwardly on channel estimation.
    - When it is expected to adapt to exceptionally particular channels, the equalizer can be further enhanced by utilizing a multi-band equalizer, rather than per-subcarrier equalizer. The thought is to utilize the yields

of the investigation channel bank compared to nearby subcarriers of utilizing just the yield of the relating subcarrier, as these yields additionally contain some helpful power.

- Previous works have demonstrated that it may be enthusiastic, in a few circumstances, to consolidate the per-subcarrier equalizer with a pre-preparation before the examination channel bank, which along these lines is regular to all subcarriers. This is, however, just intriguing when utilizing long equalizers with exceptionally specific channels and does not have any significant bearing to the remote condition considered here. What's more, this strategy isn't appropriate to a multi-client uplink situation where the channels originating from the distinctive clients may be totally extraordinary. For these reasons, this sort of equalizers won't be considered.

As a general remark, it is more effective for every one of these tasks (evening out, remuneration of synchronization blunders) to be performed in the recurrence area (at the yield of the examination channel bank) on a for every subcarrier premise with a specific end goal to exploit the selectivity of the channel bank. This additionally takes into account some greater adaptability with respect to the multi-client situation. Whatever is left of the deliverable is organized as takes after. Channel estimation is examined in light of the utilization of pilots. Specifically, the pilot is adjusted so as to consider the specific instance of the OQAM balance. In Section 7.4, diverse sorts of equalizer are introduced and their exhibitions are assessed. The remuneration of synchronization mistakes is additionally examined in that area. Section 5 introduces a more point-by-point examination of the distinctive leveling structures in a WiMAX situation and gives a couple of conclusions in regard to the decisions to be made.

REFERENCES

1. A. Farhang, N. Marchetti, L. E. Doyle, and B. Farhang-Boroujeny, "Filter bank multicarrier for massive MIMO," in *Vehicular Technology Conference (VTC Fall), 2014 IEEE 80th*, Canada, 14–17 Sept. 2014.
2. B. Farhang-Boroujeny, "OFDM versus filter bank multicarrier," *IEEE Signal Processing Magazine*, vol. 28, no. 3, pp. 92–112, 2011.
3. C. Kim, Y. H. Yun, K. Kim, and J. Y. Seol, "Introduction to QAM-FBMC: From waveform optimization to system design," *IEEE Communications Magazine Year*, vol. 54, no. 11, pp. 1–7 2016.
4. R. Haas and J. -C. Belfiore, "A time-frequency well-localized pulse for multiple carrier transmission," *Wireless Personal Communication*, vol. 5, no.1, 1997, pp. 1–18.
5. J. R. Barry, E. A. Lee, and D. G. Messerschmitt, *Digital Communication*, 3rd ed., Springer, United States, 2004.
6. Y. H. Yun et al., "A new waveform enabling enhanced QAM-FBMC systems," in *Proceedings on International Workshop on Signal Processing Advances in Wireless Communications*, New York, USA, June 2015, pp. 116–120.
7. M. Payaró, A. Pascual-Iserte, and M. Nájar, "Performance comparison between FBMC and OFDM in MIMO systems under channel uncertainty," in *2010 European Wireless Conference (EW)*, Lucca, Italy, 2010.

8. Y. Qi and M. Al-Imari, "An enabling waveform for 5G — QAM-FBMC: Initial analysis," in *2016 IEEE Conference on Standards for Communications and Networking (CSCN)*, Berlin, Germany, 2016.

9. Maziar Nekovee et al., Millimetre-wave based mobile radio access network for fifth generation integrated communications (mmMAGIC), Proc. of European Conference on Networks and Communications (EuCNC), Paris, 2015.

10. P. Sabeti, A. Farhang, N. Marchetti, and L. Doyle, "Performance analysis of FBMC-PAM in massive MIMO," in *2016 IEEE Globecom Workshops (GC Wkshps)*, USA, 2016.

11. M. Bellanger, D. Mattera, and M. Tanda, "A filter bank multicarrier scheme running at symbol rate for future wireless systems," in *IEEE Wireless Telecommunications Symposium (WTS)*, New York, USA, 2015, pp. 1–5.

12. M. Bellanger, "FS-FBMC: An alternative scheme for filter bank based multicarrier transmission," in *IEEE 5th International Symposium on Communications Control and Signal Processing (ISCCSP)*, Roma, Italy, 2012, pp. 1–4.

13. A. Farhang, N. Marchetti, L. E. Doyle, and B. Farhang-Boroujeny, "Filter bank multicarrier for massive MIMO," in *IEEE 80th Vehicular Technology Conference (VTC Fall)*, Vancouver, Canada, 2014, pp. 1–7.

14. J. G. Andrews, S. Buzzi, W. Choi, S. V. Hanly, A. Lozano, A. C. Soong, and J. C. Zhang, "What will 5G be?" *IEEE Journal on Selected Areas in Communications*, vol. 32, no. 6, pp. 1065–1082, 2014.

15. B. Le Floch, M. Alard, and C. Berrou, "Coded orthogonal frequency division multiplex," *Proceedings of the IEEE*, vol. 83, no. 6, pp. 982–996, 1995.

16. B. Farhang-Boroujeny and C. Yuen, "Cosine modulated and offset QAM filter bank multicarrier techniques: A continuous-time prospect," *EURASIP Journal on Advances in Signal Processing*, vol. 2010, Article ID165654, 16 pages, 2010.

17. B. Farhang-Boroujeny, *Signal Processing Techniques for Software Radios*, Lulu Publishing, Morrisville, North Carolina, United States, 2011.

18. R. Chang, "High-speed multichannel data transmission with band-limited orthogonal signals," *Bell System Technical Journal*, 45, 1775–1796, 1966.

19. B. Farhang-Boroujeny and C. (George) Yuen, "Cosine modulated and offset QAM filter bank multicarrier techniques: A continuous-time prospect," *EURASIP Journal on Applied Signal Processing, Special issue on filter banks for next-generation multicarrier wireless communications*, 2010, p. 6, 2010.

20. H. S. Malvar, *Signal Processing with Lapped Transforms*, Artech House, Norwood, Mass, 1992.

21. T. Karp and N. J. Fliege, "Modified DFT Filter Banks with perfect reconstruction," *IEEE Transactions on Circuits and Systems II*, vol. 46, pp. 1404–1414, 1999.

22. P. P. Vaidyanathan, *Multirate Systems and Filter Banks*, Prentice Hall, Englewood Cliffs, NJ.

8 Recent Trends in Economic Scheduling and Emission Dispatch of Distributed Generators in Microgrids

Tanuj Mishra and Amit Kumar Singh
Lovely Professional University

Vikram Kumar Kamboj
Lovely Professional University
University of Calgary

CONTENTS

8.1 INTRODUCTION

Microgrid is an augmentation of primary network capable of energy generation, equipped for satisfying its neighborhood load demand. The microgrid can be used in islanded or isolated and grid-associated modes. A microgrid design needs to be added to the main grid to build the dependability, improve power quality, stay away from the utilization of exhausting fossil fuels, improve the specialized exhibition, and lessen the ozone-depleting substances outflows. Because of these reasons, operation, control, security constraints, and grid integration of renewable resources are a task of fundamental importance in modern power systems. Microgrid operating modes and dispatch strategies must be studied. Furthermore, as renewable energy sources are intermittent in nature, energy storage schemes are required to store the energy and retrieve the energy at times required. Parimala and Ranjan [1] proposed a simple and heuristic-based solution method for solving economic load dispatch (ELD) in the islanded mode of microgrid. The economic dispatch for different groupings

DOI: 10.1201/9781003242277-8

of conventional and nonconventional sources is performed by using the real-valued genetic algorithm (GA). Results show that the integration of renewable sources and simultaneous consideration of solar credits give more profit to the microgrid. Dey et al. [2] performed all ELD, emission dispatch, and combined economic and emission dispatch (CEED) on an islanded and renewable integrated microgrid separately using a novel recently developed whale optimization algorithm (WOA). Four various scenarios of load sharing among the DERs are studied. The results are then compared with other recently developed bioinspired algorithms to corroborate the effectiveness of the proposed technique. Raj and Bhattacharyya [3] have recently developed a WOA (whale optimization algorithm) successfully and implemented it to solve the reactive power planning problem of power systems using series and shunt types of FACTS devices. It is observed that the proposed WOA provides more accurate and reliable guidance for optimal coordination of FACTS devices with other reactive power sources present in the power network. The merit that lies with the WOA is that it has a simple structure for implementation and it has ability not to be trapped in local minima, thus exploring a wider search area. It may also be concluded that the WOA may be an effective method of optimization in the field of power system engineering. Elazim [4] developed a MBA (mine blast algorithm) optimization technique to solve ELD and CEED problems in power systems. The performance of the MBA was tested in various test cases and compared with the reported results in the recent literature. The superiority of MBA over other algorithms for settling ELD and CEED problems with valve point effect is confirmed. Therefore, MBA optimization is a promising technique for solving complicated problems in power systems. Applications of the proposed algorithm to multi-area power system integrated with wind farms and PV system are the future scope of this work. Vahedipour-Dahraei et al [5] presented a stochastic model for optimal scheduling of security-constrained unit commitment associated with DR (AC-SUCDR) actions in a residential islanded microgrid. The proposed model maximized the EP of microgrid operator taking into account the AC operation constraints. The proposed AC security-constrained unit commitment and DR scheduling problem is modeled as a mixed-integer programming problem and is solved using the CPLEX solver in the general algebraic modeling system. Mashayekh et al [6] presented a novel mixed-integer linear optimization model that determines optimal technology mix, size, placement, and associated dispatch for a multi-energy microgrid. The model satisfies microgrid's electrical and heat transfer network limitations by integrating linear power flow and heat transfer equations. It captures the efficiency gains from waste heat recovery through combined heat and power technologies, by modeling the interplay between electrical and heat sources. To ensure a secure design against generator outages, the optimization maintains sufficient reserve capacity in the system, which is dynamically allocated based on system operating conditions. Liu et al. [7] proposed a cooperative reinforcement learning algorithm for distributed economic dispatch in microgrids. Applying the learning algorithm can avoid the difficulty of stochastic modeling and high computational complexity. In the cooperative reinforcement learning algorithm, the function approximation is leveraged to deal with the large and continuous state spaces. And a diffusion strategy is incorporated to coordinate the actions of DG units and ES devices. Based on the proposed algorithm, each node in microgrids only needs to communicate with its local neighbors, without relying on any centralized controllers. Algorithm convergence is analyzed, and simulations based on real-world meteorological and load data are conducted to validate the performance of the proposed algorithm. Prashar and Arora [8] applied a number of particle swarm optimization (PSO) variants to the dynamic economic emission dispatch (DEED) problem. The DEED problem is a multi-objective optimization problem in which the goal is to optimize two conflicting objectives: cost and emissions. The PSO variants tested include the standard PSO (SPSO), the PSO with avoidance of worst locations (PSO AWL), and also a selection of different topologies including the PSO with a gradually increasing directed neighborhood (PSO GIDN). The results show that the PSO AWL outperforms the SPSO for every topology implemented. The results are also compared to the state-of-the-art genetic algorithm (NSGA-II) and multi-agent reinforcement learning (MARL). Nwulu and Xia [9] investigated energy management problem for a microgrid incorporating a demand response program. The demand response program is a game theory-based demand response program (GTDR), and the grid-connected operational mode for a microgrid is investigated. The objective is to minimize the fuel cost of conventional generators and the transaction cost for trading transferable power and at the same time maximize the grid operator DR's benefit. The optimization model has a scheduling interval of 24 hours and determines the optimal customer power curtailed, optimal customer incentive, optimal power generation schedule for the conventional generators, and optimal power to be transferred between the main grid and microgrid. The Advanced Interactive Multidimensional Modeling System (AIMMS) is used to solve the developed model, and the obtained results indicate that incorporating DR programs into the energy management of microgrid problem is helpful and introduces optimality at both the supply and demand sides of the microgrid. Mirjalili and Lewis [10] presented a new swarm-based optimization algorithm inspired by the hunting behavior of humpback whales. The proposed method (named as WOA, whale optimization algorithm) included three operators to simulate the search for prey, encircling prey, and bubble-net foraging behavior of humpback whales. An extensive study was conducted on 29 mathematical benchmark functions to analyze exploration, exploitation, local optima avoidance, and convergence behavior of the proposed algorithm. WOA was found to be enough competitive with other state-of-the-art metaheuristic methods (Table 8.1).

8.2 MICROGRID ARCHITECTURE

Economic load dispatch (ELD) is the key problem related to the operation of grid-connected or islanded microgrids. The goal of the economic load dispatch problem is to share the output power of the running generation sources so as to provide the load demand satisfying the generator constraints at a minimized fuel cost. Accordingly, numerous optimization techniques are implemented to solve complex and convex ELD problems. Some of them include the participation factor methods, the gradient methods, the linear methods, and Newtonian methods. These methods may be simple, but they converge toward the solution very slowly. Furthermore, the harm caused to the environment by the release of the toxic gases in the atmosphere has gained much attention in the last few decades by the utility companies. They are

TABLE 8.1

Comparative Analysis of the Literature Work

Ref. No.	Description	Work Done	Technique Used	Future Work Scope
[11]	Energy and operation management of a microgrid	Minimization of the microgrid cost comprising of a microturbine fuel cell, PV, wind turbine, and battery storage	PSO (particle swarm optimization)	It can serve as a useful decision-making support tool for microgrid operators and help to determine how the input random variables affect the statistical characteristics of the EOM
[12]	Consideration of uncertainties in load demand, market prices, and the available electrical power of wind farms and photovoltaic systems	The forecasted value of PV and wind turbine and also the real-time market prices were considered while minimizing the microgrid cost	Point estimate method and self-adaptive gravitational search algorithm	It could find robust, reliable, and high-quality solutions in a satisfactory simulation time for energy management problems
[13]	Annual economic/ environmentally optimal operation of a low-voltage microgrid with DG capabilities (FC, MT and PV units)	Minimization of the microgrid cost involving penetrable PV, microturbine, and fuel cell	Harmony search algorithm (HSA)	Electric grids will be able to be combined with gas and hydrogen infrastructures. Such a progress will increase the penetration of fuel cells even more
[6]	Security-constrained design of isolated multi-energy microgrids	Ensures a secure design against generator outages; the optimization maintains sufficient reserve capacity in the system, which is dynamically allocated based on system operating conditions	Mixed-integer linear optimization model	Future research will focus on integrating network design (cable/pipe placement and sizing) into the model

bound to maintain certain levels regarding the release of harmful gases such as carbon dioxide (CO_2), carbon monoxide (CO), and sulfur dioxide (SO_2). The emission of these harmful gases can be reduced by installing more efficient and cleaner generators that consume less fuel, updating the control equipment, and emission dispatch. Emission dispatch was first performed to minimize the emission of nitrous oxide (NO_x) gases, but the corresponding ELD proved to be costlier. The economic emission dispatch (EED) idea was brought to find a compromised solution between the cost and emission levels. A fuzzy interval optimization approach was used to solving the EED problem with uncertain parameters in the constraints. Various algorithms were applied to solve emission problems subject to various constraints. Evolvement of soft computing tools, which are not restricted by complexity of system models, inspired the research workers to apply them in the field of power system optimization. Due to the limited/lack of access to the main grid, microgrids have been the only way out for rural/remote and off-grid communities for a long time. Usually, these off-grid communities relied on diesel generation to supply their loads, in spite of the higher fuel prices, but nowadays with increased global environmental concerns and incentives for a transformation of these diesel-based systems into renewable-based microgrids, changes can be observed. This transition to resources with high variability and uncertainty, such as wind and photovoltaic systems, requires new microgrid planning tools to ensure the security of supply of these isolated systems. A comprehensive microgrid investment and planning optimization must address (i) power generation mix selection; (ii) resource sizing and allocation; (iii) operation scheduling; and (iv) interplay between electricity, cooling, and heating loops in the microgrid to take full advantage of excess heat. (v) Moreover, in the context of remote/isolated microgrids, accounting for security of supply constraints in the design and operation is needed.

8.3   ECONOMIC DISPATCH AND EMISSION

The objective of ED problem is to simultaneously minimize the total generation cost (F_T) and to meet the load demand of a power system over some appropriate period while satisfying various constraints. The objective function is

$$F_T = \min\left(\sum_{i=1}^{n} F_i(P_{Gi})\right) = \min\left(\sum_{i=1}^{n} A_i P_{Gi}^2 + B_i P_{Gi} + C_i\right) \tag{8.1}$$

where P_{Gi} is the power generation of unit i; $F_i(P_{Gi})$ the generation cost function; and A_i, B_i, and C_i the cost coefficients of ith generator. There are two constraints considered in the problem, i.e., the generation capacity of each generator and the power balance of the entire power system.

- **Constraint 1**: Generation capacity constraint
  For normal system operations, the real power output of each generator is restricted by lower and upper bounds as follows:

$$P_{Gi}^{\min} \leq P_{Gi} \leq P_{Gi}^{\max}$$

The minimum and maximum powers are generated by ith generator, respectively.

- **Constraint 2**: Power balance constraint

The total power generation must cover the total demand P_D and the real power loss in transmission lines P_L. This relation can be expressed as follows:

$$\sum_{i=1}^{n} P_{Gi} = P_D + P_L \tag{8.2}$$

Here, a reduction is applied to model transmission losses as a function of the generators output through Kron's loss coefficients. The Kron's loss formula can be expressed as follows:

$$P_L = \sum_{i=1}^{n}\sum_{j=1}^{n} P_{Gi}B_{ij}P_{Gj} + \sum_{i=1}^{n} B_{oi}P_{Gi} + B_{oo} \tag{8.3}$$

where B_{ij}, B_{oi}, and B_{oo} are the transmission network power loss B-coefficients.

Various pollutants such as carbon dioxide, sulfur dioxide, and nitrogen oxide are released as a result of the operation of the diesel generator, gas generator, and CHP. Reduction in these pollutants is compulsory for every generating unit. To achieve this goal, new criteria are included in the formulation of the emission dispatch problem as follows:

$$E_T = \sum_{i=1}^{n}\left(x_i \cdot P_i^2 + y_i P_i + z_i\right) \tag{8.4}$$

where E_T = the total emission value, x_i = the emission coefficient of ith generator in [kg/MWh], y_i = the emission coefficient of ith generator in [kg/MWh], and z_i = the emission coefficient of ith generator in [kg/h]. There are certain cases that can be considered in microgrids where the generations by different distributed generators can be selected assuming different scenarios of renewable energy sources (Figure 8.1).

8.4 CONCLUSIONS

It is proposed here that microgrids play a vital role in improving the power system and making it smart. Also, the major issue related to the emission of harmful gases, reliability, energy security, and economic load dispatch can be taken care of efficiently by implementing various optimization techniques. A lot of improvement is still required as it is a continuous process and needs to manage certain area as new load demands and changes in energy requirements are happening day to day as the technology is growing. There is a lot of scope in incorporating electric vehicle system in microgrids and energy systems used. The use of renewable energy resources and

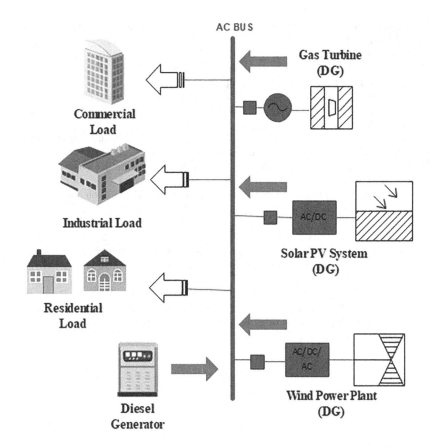

FIGURE 8.1   Architecture of an islanded microgrid.

their maintenance and installation is also important, and its cost must be considered while calculating the total cost of the microgrid.

REFERENCES

1. N. Parimala and R. K. Ranjan, "Soft computing in data analytics," *Proceedings of International Conference on SCDA 2018*, Andhra Pradesh, India, vol. 758, pp. 1–848, 2019.
2. B. Dey, S. K. Roy, and B. Bhattacharyya, "Solving multi-objective economic emission dispatch of a renewable integrated microgrid using latest bio-inspired algorithms," *Eng. Sci. Technol. Int. J.*, vol. 22, no. 1, pp. 55–66, 2019.
3. S. Raj and B. Bhattacharyya, "SC," *Swarm Evol. Comput. BASE DATA*, 2018.
4. E. S. Ali and S. M. A. Elazim, "Mine blast algorithm for environmental economic load dispatch with valve loading effect," *Neural Comput. Appl.*, vol. 30, no. 1, pp. 261–270, 2018.
5. M. Vahedipour-Dahraei, H. R. Najafi, A. Anvari-Moghaddam, and J. M. Guerrero, "Security-constrained unit commitment in AC microgrids considering stochastic price-based demand response and renewable generation," *Int. Trans. Electr. Energy Syst.*, vol. 28, no. 9, p. e2596, 2018.

6. S. Mashayekh et al., "Security-constrained design of isolated multi-energy microgrids," *IEEE Trans. Power Syst.*, vol. 33, no. 3, pp. 2452–2462, 2018.

7. W. Liu, P. Zhuang, H. Liang, J. Peng, and Z. Huang, "Distributed economic dispatch in microgrids based on cooperative reinforcement learning," *IEEE Trans. Neural Networks Learn. Syst.*, vol. 29, no. 6, pp. 2192–2203, 2018.

8. D. Prashar and K. Arora, "Design of two area load frequency control power system under unilateral contract with the help of conventional controller," *IJICTDC*, 5, pp. 22–27, 2020.

9. N. I. Nwulu and X. Xia, "Optimal dispatch for a microgrid incorporating renewables and demand response," *Renew. Energy*, vol. 101, pp. 16–28, 2017.

10. S. Mirjalili and A. Lewis, "Advances in engineering software the whale optimization algorithm," *Adv. Eng. Softw.*, vol. 95, pp. 51–67, 2016.

11. J. Radosavljević, M. Jevtić, and D. Klimenta, "Energy and operation management of a microgrid using particle swarm optimization," *Eng. Optim.*, vol. 48, no. 5, pp. 811–830, 2016.

12. T. Niknam, F. Golestaneh, and A. Malekpour, "Probabilistic energy and operation management of a microgrid containing wind / photovoltaic / fuel cell generation and energy storage devices based on point estimate method and self-adaptive gravitational search algorithm," *Energy*, vol. 43, pp. 427–437, 2012.

13. A. G. Anastasiadis, S. A. Konstantinopoulos, G. P. Kondylis, G. A. Vokas, and P. Papageorgas, "Effect of fuel cell units in economic and environmental dispatch of a Microgrid with penetration of photovoltaic and micro turbine units," *Int. J. Hydrogen Energy*, vol. 42, pp. 6–13, 2016.

9 Forecasting of Tensile–Shear Strength of JSC 590RN Low-Carbon Steel Spot Welds Using Taguchi Technique Used in Electrical Grids

Abhishek Tyagi, Gaurav Kumar, and Mukesh Kumar
Dr A P J Abdul Kalam Technical University

Krishan Arora
Lovely Professional University

CONTENTS

9.1   INTRODUCTION

Spot welding is the earliest resistance welding method applied in the manufacturing world for a long time due to more considerable electrical resistance and lower thermal conductivity (λ) [1]. Due to several advantages such as low cost of joining, easy process, and high productivity, RSW is used in many industries. If the level of current (I) and nugget size increase, then the joint's strength decreases [2]. In the RSW process, mainly aluminum and low-carbon steel sheets are utilized for coalescence. Various essential parameters manage the spot welding procedure, i.e., diameter of the electrode, electrode pressure, weld current, weld time, and electrode and sheet materials. The best qualitative parameters of spot welding are the size of nugget diameter, the value of heat-affected zone, and tensile load. The outcomes of such output characteristics depend on the RSW input factors [3]. During the RSW process, a large amount of current is needed for melting and joining the metal sheet, which will cause to warm the electrode and reduce the continuance of the electrode cover. So, to reduce this issue of extreme electrode heating, water is utilized as a coolant [4]. RSW is a regular coalescence process in the automobile and manufacturing industry, and it supersedes the other. In the RSW process, steel and other light sheet alloys face many difficulties. These difficulties are extra heating, flushing, and melting [5]. It is the metal sheet amalgamation procedure used in many areas, such as automobile, aircraft, railways, spacecraft fabrication, and domestic appliances. There are three to six thousand spot welds used in any automotive body, which indicates the importance of RSW [6].

Tutar et al. [7] selected an aluminum compound (AA3003-H12) sheet to optimize the FSSW parameters by applying the L9 Taguchi technique. They selected plunge time (seconds), dwell time (seconds), and rotation speed (revolutions per minute) as the three process parameters to determine the optimization outcomes of T–S strength and found that the plunge depth was the essential parameter among others. Ghogale and Patil [8] used Taguchi methodology to improve the quality of MIG welding. This investigation aimed to find the excellent value of penetration by taking the main process factors of MIG welding, such as current, voltage, and flow of gas. Patel [9] also used the Taguchi proposal to regulate the effect of input factors on T–S load. ANOVA is conducted to examine the outcomes, and it was found that the current contributes more than other parameters. Muthu [10] experimented on the SS 316 sheet to improve the RSW parameters by Taguchi L_{27} design. As reported in this research, it has been found that the electrode diameter is the very crucial factor of T–S strength relative to other parameters. Dhawale and Name [11] carried out an investigation to predict the tensile and shear strengths of RSW by applying the Taguchi L_{27} design. Plan input factors are engaged to investigate tensile–shear load. Singh et al. [12] presented a review paper to discuss the contribution of the Taguchi and GRA methods in the improvement of milling input factors. They performed shoulder milling on AA6063 T6 by applying Taguchi method with a GRA to optimize the process constraints. Kumar et al. [13] experimented with optimizing the end milling on SS 304 by the Taguchi method. Muhammad et al. [14] presented an investigation to optimize the input factor of RSW to increase the weld quality. They took nugget size and HAZ as quality parameters and applied the Taguchi L_9 OA technique. Ghazali et al. [15] investigated the optimization of the RSW factor by using the Taguchi L_9 design. They took three input parameters, i.e., electrode force, weld time, and weld current, to determine quality characteristics, i.e., tensile–shear strength and nugget diameter. Their research indicates that the weld current is the supreme parameter that affects the load strength and nugget diameter. Sao and Banchor [16] experimented with optimizing the spot welding factor by applying the GRA method. They concluded that the current is the very vital parameter to optimize weld characteristics. Goel et al. [17,18] done friction stir welding to join two different materials, namely AA-7475-T761 and AISI 304, using the L_8 Taguchi technique. They used various parameters to determine the highest value of UTS and found rotational tool speed as the most contributing factor.

The research paper analysis indicates that the RSW issue of mild steel (JSC 590RN) has many problems: electrode deposit on the weld, ruptures, porosities, weld area, and challenges in finding out the standard of welds such as T–S load. The difficulty related to the coalescence of JSC 590RN by the KUKA robot is the primary objective for selecting the job metals. The effect of input factors of RSW such as electrode diameter, electrode force, weld current, and weld time on T–S load was observed. The Taguchi L_{18} design was enforced for the optimization of spot welding.

9.2 MATERIALS AND METHODS

9.2.1 WORKPIECE DESIGN

CR low-carbon steel (JSC 590RN) sheets with a thickness of 1.2 mm were exercised to weld the workpiece. 160 mm × 50 mm × 1.2 mm is the size of one sample with a lapping of forty millimeters on other samples having a similar size, as illustrated in Figure 9.1. The present material sheet (description – JSC 590RN, commodity – cold-rolled sheet, Bureau of Indian Standards grade – IS 513 (Part 2), ISC 590LA) is mainly used for forming automotive equipment, mainly used in cross-member dash (upper side) of the automobile.

For the chemical composition test, a small part is removed from the JSC 590RN (mild steel) sheet. The constitution of the sheet material is illustrated in Table 9.1.

9.2.2 EXPERIMENTAL PROCEDURE

Four input factors are selected in the present work. Their values are under the Taguchi L_{18} technique design illustrated in Table 9.2.

All experimental trials are conducted in VEEGEE Industrial Enterprises, Faridabad, by a KUKA robot (Figure 9.2).

All 18 trials were done according to the Taguchi L_{18} method. The tensile–shear strength of all the samples is determined using a UTM of version TUE-C-400

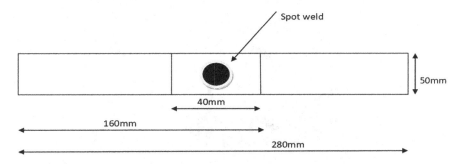

FIGURE 9.1 Lap joint for spot welding.

TABLE 9.1
JSC 590RN Chemical Composition

Content	Value %
C	0.144
Mn	1.37
Si	0.240
P	0.013
S	0.003

TABLE 9.2
Spot Welding Factors at Various Scales

Symbol	Factors	Scale 1	Scale 2	Scale 3
A	Electrode diameter/length (mm)	13/23	16/23	–
B	Electrode force (N)	2,650	2,750	2,850
C	Weld current (A)	9,700	9,800	9,900
D	Welding time (cycle)	10	12	14

FIGURE 9.2   KUKA robot.

FIGURE 9.3   UTM TUE-C-400 carbon steel model.

(illustrated in Figure 9.3). The failure of the specimen before and after testing is also shown in Figure 9.4.

The working prerequisite of this research is illustrated in Table 9.3.

9.3 RESULTS AND DISCUSSION

9.3.1 *S/N* RATIO

Taguchi methods require analyzing the data via a signal-to-noise ratio that will provide two advantages; it helps guide the excellent scale depending on the slight variation throughout the mean value, which is closest to the target value. The ratio of "signal" to "noise" means acceptable value and non-acceptable value. Based on quality engineering (QE), the features are categorized into three, i.e., higher is better, lower is better, and nominal is better. T–S strength and nugget diameter need higher values. Likewise, HAZ needs a lower value. The descriptions of this equation can be found in Ref. [19]. In this investigation, for T–S strength the category higher is better is chosen. Equation (9.1) is used for the maximization of T–S strength.

$$S/N = -10 \times \log\left(\sum\left(1/Z^2\right)/n\right) \qquad (9.1)$$

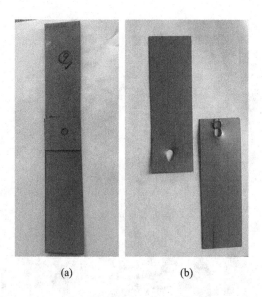

(a)   (b)

FIGURE 9.4   Failure of the specimen: (a) before testing and (b) after Testing.

TABLE 9.3
Spot Welding Prerequisite

Prerequisite	Specification
Materials	Low-carbon steel (JSC 590RN)
Size	160 mm × 50 mm × 1.2 mm
Welding instrument	KUKA robot
Measuring appliances	UTM (TUE-C-400)

where n is the number of tests and Z is the observed experimental data.

The Taguchi L_{18} experimental method was used for all 18 experimental trials. The T–S strength of all the weld joints can be determined by the UTM of model TUE-C-400. The measured values of T–S strength and their S/N ratio are illustrated in Table 9.4.

The total average of S/N ratio = 24.21.

From Figure 9.5, it can be noticed that the S/N value of electrode diameter is reducing on raising their levels. However, the S/N values of electrode force, weld current, and weld time raised their second levels. The dashed line shows the average of the S/N value, i.e., 24.21. Table 9.5 illustrates that the distinction between the maximum and minimum values of the S/N level of parameters C (0.25) was the maximum, followed by the parameters D (0.15), A (−1.01), and B (−1.32). Also, from Figure 9.5, it can be observed that the signal-to-noise ratio values peaked at the 1st level of electrode diameter, 2nd level of electrode force, 2nd level of weld current, and also 3rd level of welding time. So $A_1B_2C_2D_3$ is the most acceptable conjunction of quality features for spot welding procedures.

TABLE 9.4

Taguchi OA with T–S Strength and Its *S/N* Ratio

Run No.	A	B	C	D	T–S (kN)	S/N Ratio
1	1	1	1	1	11.31	21.07
2	1	1	2	2	19.91	25.98
3	1	1	3	3	21.09	26.48
4	1	2	1	1	14.41	23.17
5	1	2	2	2	24.47	27.77
6	1	2	3	3	23.93	27.58
7	1	3	1	2	13.50	22.61
8	1	3	2	3	16.58	24.39
9	1	3	3	1	14.77	23.39
10	2	1	1	3	12.23	21.75
11	2	1	2	1	15.75	23.95
12	2	1	3	2	13.35	22.51
13	2	2	1	2	10.46	20.39
14	2	2	2	3	17.20	24.71
15	2	2	3	1	20.09	26.06
16	2	3	1	3	15.76	23.95
17	2	3	2	1	18.65	25.41
18	2	3	3	2	17.13	24.68

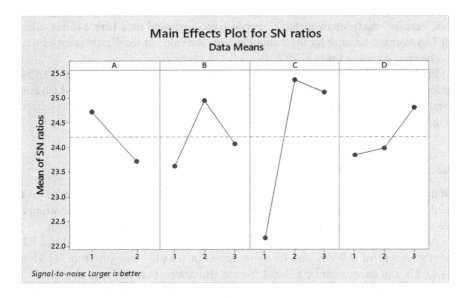

FIGURE 9.5   Average of *S/N* ratio.

TABLE 9.5
S/N Ratio Response Table

Symbol	Input Parameter	Scale 1	Scale 2	Scale 3	Max–Min
A	Electrode diameter	24.72	23.71	–	−1.01
B	Electrode force	23.62	24.95	24.07	−1.32
C	Welding current	22.16	25.36887	25.11539	0.25
D	Weld time	23.84	23.98895	24.80992	0.15

9.3.2 ANALYSIS OF VARIANCE

ANOVA incorporates to judge the essential factors on the quality features of spot welding. A statistical tool, F-test, was used to find the effect of these input factors on quality features. When its value is more significant than four, this implies that modification in welding factors has more impact on quality features [20]. In ANOVA, the null hypothesis states that no difference in means exists. The research or alternative hypothesis is that the means are not all equal, and it is frequently expressed in words rather than mathematical symbols. The study hypothesis captures any variation in means, which includes situations where all four means are unequal, one is different from the other three, two are different, and so on. As illustrated above, the alternative hypothesis captures all possible scenarios other than the null hypothesis's specification of equality of all means.

The distinguishing coefficient (α) lies between 0 and 1, i.e., $\alpha = 0.5$, and then the confidence level of ANOVA was 95%.

The outcomes of ANOVA illustrated in Table 9.6 verify that out of all input parameters of spot welding, welding current contributes more in contrast to other input factors. Besides, electrode diameter, electrode force, and weld time have a minor effect on T–S strength because for these input factors, the value of the F ratio is lesser than 4. The error measured was 31.79%.

From ANOVA, the most contributing factor is the welding current, i.e., 50.67%, followed by electrode force (F), electrode diameter (d), and weld time (t), which contribute 7.21%, 6.01%, and 4.32%, respectively. This indicates that the welding current is the most significant process factor compared to other process factors.

9.3.3 CONFIRMATION TEST

To determine the optimization result of this research, the Taguchi L_{18} method was used. Table 9.7 illustrates that the investigation outcomes of tensile–shear strength are in most excellent agreement with the prediction of the investigation design.

According to Table 9.7, the signal-to-noise ratio changed from 27.20 to 27.77 (an increment of 2.06%) due to an increase in tensile strength from 22.91 to 24.47 kN (an increment of 6.38%), which illustrates that the best combination ($A_1B_2C_2D_3$) is good enough for maximizing the tensile–shear strength (T–S strength).

TABLE 9.6

Outcomes of ANOVA

Input Factor	SS	DoF	MS	F Ratio	Percentage Contribution
Electrode diameter	4.55	1	4.55	1.89	6.01
Electrode force	5.45	2	2.72	1.13	7.21
Welding current	38.30	2	19.15	7.97	50.67
Weld time	3.27	2	1.63	0.68	4.32
Error	24.02	10	2.40		31.79
Total	75.58	17			100.00

TABLE 9.7

Outcomes of Confirmation Experiment

	Predication	Experiment	% Change
Level	$A_1B_2C_2D_3$	$A_1B_2C_2D_3$	
T–S (kN)	22.91	24.47	6.38
S/N ratio	27.20	27.77	2.06

9.4 CONCLUSIONS

On observing the result outcomes, various conclusions can be made.

- The best combination for the T–S strength of the spot welding process is $A_1B_2C_2D_3$.
- From ANOVA, the most contributing factor is welding current, i.e., 50.67%, followed by electrode force (F), electrode diameter (d), and weld time (t), which contribute 7.21%, 6.01%, and 4.32%, respectively. This indicates that the welding current is the most significant process factor in contrast to other input factors. According to this technique, the outcomes can be enhanced up to 2.06%.

REFERENCES

1. Raut M., Achwal V. (2014). Optimization of spot-welding process parameters for maximum tensile strength. *International Journal of Mechanical Engineering and Robotics Research*, 3(4), 507–517.
2. Mat Y.N.A., Alisibramulisi A., Salleh Z., Ghazali F.A., Pawan A. (2020). Optimization of resistance spot welding (RSW) parameters by using Taguchi method. *International Journal of Innovative Technology and Exploring Engineering*, 9(3), 2795–2800.
3. Thakur A.G., Rao T.E., Mukhedkar M.S., Nandedkar V.M. (2010). Application of Taguchi method for resistance spot welding of galvanized steel. *ARPN Journal of Engineering and Applied Sciences*, 5(11), 22–26.
4. Rawal M. R., Kolhapure R. R., Sutar S. S., Shinde V. D. (2016). Optimization of resistance spot welding of 304 steel using GRA. *International Journal*, 3(9), 492–499.

5. Prashar D., Arora K. (2020). Design of two area load frequency control power system under unilateral contract with the help of conventional controller. *IJICTDC*, 5, 22–27.

6. Pandey A. K., Khan M. I., Moeed K. M. (2013). Optimization of resistance spot welding parameters using Taguchi method. *International Journal of Engineering Science and Technology*, 5(2), 234–241.

7. Tutar M., Aydin H., Yuce C., Yavuz N., Bayram A. (2014). The optimisation of process parameters for friction stir spot-welded AA3003-H12 aluminium alloy using a Taguchi orthogonal array. *Materials & Design*, 63, 789–797.

8. Ghogale M. M., Patil S. A. (2013). Optimisation of process parameters of MIG welding to improve quality of weld by using Taguchi methodology. *International Journal of En gineering Research & Technology*, 2(12), 3677–3685.

9. Patel K., Arora K., Kaur P. (2019). "Power monitoring system in solar power plant using LabVIEW," *2019 2nd International Conference on Intelligent Computing, Instrumentation and Control Technologies (ICICICT)*, pp. 1011–1015, doi:10.1109/ICICICT46008.2019.8993249.

10. Muthu P. (2019). Optimization of the process parameters of resistance spot welding of AISI 316l sheets using Taguchi method. *Mechanics and Mechanical Engineering*, 23(-1), 64–69.

11. Dhawale P. A., Name B. R. (2019, December). Prediction of weld strength by parametric optimization of resistance spot welding using Taguchi method. In *AIP Conference Proceedings* (Vol. 2200, No. 1, p. 020087). AIP Publishing LLC.

12. Singh O. P., Kumar G., Kumar M. (2019). Multi performance optimization of shoulder milling process parameters of AA6063 T6 aluminium alloy by Taguchi Based GRA. *International Journal of Innovative Technology and Exploring Engineering*, 8, 420–425.

13. Kumar G., Kumar M., Tomer A. (2021). Optimization of end milling machining parameters of SS 304 by Taguchi technique. In Arunesh Chandra (ed.), *Recent Advances in Mechanical Engineering* (pp. 683–689). Springer, Singapore.

14. Muhammad N., Manurung Y. H., Hafidzi M., Abas S. K., Tham G., Haruman E. (2012). Optimization and modeling of spot-welding parameters with simultaneous multiple response consideration using multi-objective Taguchi method and RSM. *Journal of Mechanical Science and Technology*, 26(8), 2365–2370.

15. Ghazali F. A., Manurung Y. H., Mohamed M. A. (2014). Multi-response optimization using Taguchi method of resistance spot welding parameters. In Amir Khalid, Bukhari Manshoor, Erween Abdul Rahim, Waluyo Adi Siswanto and Kamil Abdullah (eds) *Applied Mechanics and Materials* (Vol. 660, pp. 120–124). Trans Tech Publications Ltd.

16. Sao M., Banchor R. (2016). Optimization of resistance spot weld parameters using grey relational analysis. *International Journal for Scientific Research and Development*, 4(10), 375–378.

17. Goel P., Saxena A. K., Wahid M. A., Rathore S., Sharma N., Mishra K. M. (2020). Influence of Friction Stir Welding parameters on mechanical properties of dissimilar AA7475 to AISI 304. *IOP Conference Series: Materials Science and Engineering*, 802(1), 012010.

18. Goel P., Mohd A.W., Sharma N., Siddiquee A.N., Khan Z.A. (2019). Effect of Welding parameters in friction stir welding of stainless steel and aluminum alloy. In Shanker K., Shankar R., Sindhwani R. (eds) *Advances in Industrial and Production Engineering* (pp. 815–823). Springer, Singapore.

19. Siddiquee A. N., Khan Z. A., Mallick Z. (2010). Grey relational analysis coupled with principal component analysis for optimisation design of the process parameters in in-feed centreless cylindrical grinding. *The International Journal of Advanced Manufacturing Technology*, 46(9), 983–992.
20. Siddiquee A. N., Khan Z. A., Goel P., Kumar M., Agarwal G., Khan N. Z. (2014). Optimization of deep drilling process parameters of AISI 321 steel using Taguchi method. *Procedia Materials Science*, 6, 1217–1225.

10 Experimental Analysis of Surface Integrity of Machined Stainless Steel (SS-304) by Taguchi Method Coupled with GRA Used in Electrical Grids

Gaurav Kumar
Vidya College of Engineering

Kapil Kumar Goyal
Dr. B. R. Ambedkar National Institute of Technology

Mukesh Kumar
Vidya College of Engineering

Krishan Arora
Lovely Professional University

CONTENTS

DOI: 10.1201/9781003242277-10

10.1   INTRODUCTION

Milling is a basic machining technology commonly used in machine shops and industries today for cutting parts to exact shapes and sizes that are used in combination with
other components in applications such as aerospace, automobile, die, and equipment
design [1]. End milling is essential among the various milling operations because it can
produce complicated geometric surfaces with reasonable precision and surface smooth-
ness. It can also produce several combinations when used in conjunction with the milling cutter [2]. A material's safety, dependability, and life cycle costs can be increased
by machining it to suitable specified service requirements [3]. The Taguchi-based grey
relational analysis is utilized to establish machining parameters and optimize cutting
parameters for stainless steel 304 to determine the optimal parameter combination.

Hamdan et al. [4] applied Taguchi method to characterize the performance of a
coated carbide tool throughout the machining of stainless steel. Prashar and Arora [5]
studied the performance constraints in turning SS-304 with dry conditions. Uncoated
cemented carbide inserts were used. It was based on grey relational analysis. Kumar
et al. [6] worked on SS-304 to improve the surface quality by applying Taguchi method.
Kumar et al. [7] improved the SS-321's multi-performance machining properties. To
optimize the drilling settings, the grey relational analysis with Taguchi method was
utilized. According to Karnwal et al. [8], the Taguchi technique paired with GRA can
be utilized to study diesel engine performance. Patel et al. [9] optimized the turning
constraints through Taguchi method. It was discovered by Ho and Lin [10] that Taguchi
process data might be utilized in conjunction with GRA to determine the impacting
order of parameters. Singh et al. [11,12] improved the surface integrity of AA6063T6
aluminum alloy through shoulder milling using Taguchi L_{18} coupled with GRA.

As per the above-discussed literature, it can be concluded that grey relational analysis along with Taguchi is the most useful technique for the optimization of parameters during end milling of stainless steel. In this study, process parameters under
consideration are coolant, feed rate, depth of cut, and speed. Surface roughness (SR),
material removal rate (MRR), and microhardness are opted as output parameters.

10.2   EXPERIMENTAL PARTICULARS

10.2.1   MATERIALS

Stainless steel 304 was chosen as the work piece material for this study because
of its high application in industries. The sample is in the dimension of 80 mm × 50

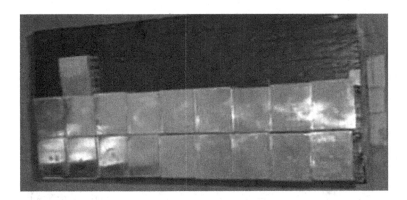

FIGURE 10.1   SS-304 pieces.

TABLE 10.1
(% Weight) of Elements in SS-304

Weight	70.37	0.042	0.345	1.76	<0.005	18.38	<0.01	0.361	0.049	0.066
Name of element	Fe	C	Si	Mn	S	Cr	Al	Cu	Nb	V

mm × 10 mm, as shown in Figure 10.1. Chemical spectrometry is used to determine the chemical composition, as shown in Table 10.1.

10.2.2   EXPERIMENTAL SETUP

A solid carbide tool was used for machining on SS-304 pieces through a vertical milling machine (VMM). Figure 10.2 depicts an overview of the process used in the current study.

10.2.3   EXPERIMENTAL DESIGN

The different machining parameters are shown in Table10.2, where each parameter has three levels except for coolant, which has two levels.

10.2.4   RECORDING OF RESPONSE CHARACTERISTICS

Figure 10.3 shows the surface roughness measurement device used to measure the roughness values of each sample. After metallographic finishing, the sample underwent Vickers hardness test in this study (see Figure 10.4). A constant force of 20 kg was applied throughout the testing, and all microhardness values were measured and noted. The MRR is calculated as follows:

$$\text{Material Removal Rate} = \frac{w \times f \times d}{60} \text{ mm}^3/\text{s} \qquad (10.1)$$

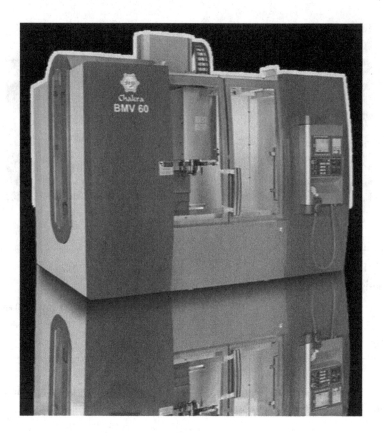

FIGURE 10.2   Experimental setup for end milling operation.

TABLE 10.2
Machining Range of the Parameters

Notation	A	B	C	D
Parameter	Coolant	Feed (mm/min)	Depth of Cut (mm)	Speed (rpm)
1	ON	1,500	0.2	1,500
2	OFF	2,000	0.3	2,500
3	–	2,500	0.4	3,500

where f, d, and w are represented as feed-in mm/min; depth of cut in mm; and width of block in mm. Each sample width is constant, i.e., 10 mm.

Each machined sample's maximum shear strain areas were measured, and the surface microstructure was examined using 600× micrographs. Table 10.3 illustrates the values of SR, MRR, and MH.

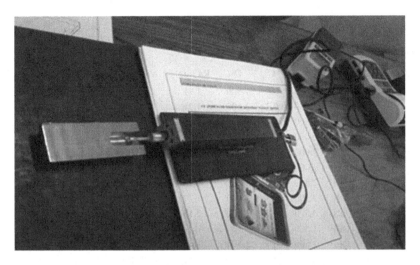

FIGURE 10.3 Surface roughness tester.

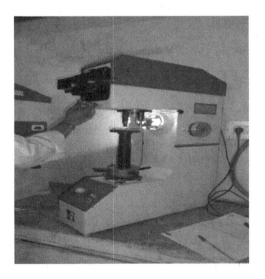

FIGURE 10.4 Vickers hardness tester.

10.3 METHOD

10.3.1 TAGUCHI TECHNIQUE (SIGNAL-TO-NOISE (S/N) RATIO)

Taguchi's method is the most effective research technique for refining quality while lowering costs. The Taguchi technique employs a unique OA design to examine the all assigned parameters. The grey relational analysis regulates the

TABLE 10.3
Experimented Responses

Experiments	A	B	C	D	Ra (μm)	MRR (mm³/s)	Microhardness (HV)
1	ON	1,500	0.2	1,500	1.91	50	220
2	ON	1,500	0.3	2,500	1.21	75	237.95
3	ON	1,500	0.4	3,500	1.75	100	196.64
4	ON	2,000	0.2	1,500	2.35	66.66	216
5	ON	2,000	0.3	2,500	2.3	100	173
6	ON	2,000	0.4	3,500	2.18	133.33	193.12
7	ON	2,500	0.2	2,500	1.96	83.33	203.88
8	ON	2,500	0.3	3,500	1.79	125	176.57
9	ON	2,500	0.4	1,500	2.15	166.66	244.9
10	OFF	1,500	0.2	3,500	1.5	50	153.52
11	OFF	1,500	0.3	1,500	1.45	75	202.58
12	OFF	1,500	0.4	2,500	1.38	100	162.1
13	OFF	2,000	0.2	2,500	1.41	66.66	184.78
14	OFF	2,000	0.3	3,500	1.08	100	188.54
15	OFF	2,000	0.4	1,500	1.81	133.33	184
16	OFF	2,500	0.2	3,500	1.83	83.33	161.21
17	OFF	2,500	0.3	1,500	1.98	125	239.74
18	OFF	2,500	0.4	2,500	1.61	166.66	284

ideal degree of numerous regulated elements at the same time. Three S/N ratios are commonly utilized, and their choice is based on reaction characteristics. The criterion of "lower is better" will be used for roughness, whereas for microhardness and MRR, "higher is better" will be used. The following are the formulas for all three S/N ratios [13].

10.3.2 STEPS OF GREY RELATIONAL ANALYSIS (GRA)

10.3.2.1 Normalization

The initial stage in GRA is the normalization of the trial data between 0 and 1. Normalization facts are explained in Refs. [14–16].

10.3.2.2 Analysis of GRG (Grey Relational Grade) and GRC (Grey Relational Coefficient)

The following formulas are used for grey relational analysis [11].

$$\xi_i(l) = \frac{\nabla_{min} + \xi \cdot \nabla_{max}}{\nabla_{0i}(l) + \xi \cdot \nabla_{max}}$$ (10.2)

$$\Delta_{0i}(l) = \left\| z_0^*(l) - z_i^*(l) \right\|$$ (10.3)

$$\Delta_{max} = \forall\overset{max}{j} \in i \ \forall\overset{min}{l} \left\| z_0^*(l) - z_i^*(l) \right\| \tag{10.4}$$

$$\Delta_{min} = \forall\overset{max}{j} \in i \ \forall\overset{min}{l} \left\| z_0^*(l) - z_i^*(l) \right\| \tag{10.5}$$

$$\gamma_i = \frac{1}{m} \sum_{l=1}^{n} \xi_i(l) \tag{10.6}$$

$$\gamma_i = \frac{1}{m} \sum_{l=1}^{m} w_l \xi_i(l) \sum_{l=1}^{m} w_l \tag{10.7}$$

The value of distinguishing coefficient (ζ) was 0.5.

10.4 OUTCOMES AND EXPLANATION

10.4.1 BEST SUITABLE COMBINATION OF PARAMETERS

Table 10.4 shows all the different values of GRC with GRG.

TABLE 10.4
Evaluated GRC and Grade Values

Expt. No.	GRC			GRG	Order
	SR	MRR	MH		
1	0.6522	0.3333	0.5464	0.5106	10
2	0.3693	0.4298	0.6348	0.4780	12
3	0.5687	0.5410	0.4555	0.5217	9
4	1.0000	0.0000	0.5291	0.5097	11
5	0.9476	0.5410	0.3829	0.6238	5
6	0.8381	0.7296	0.4437	0.6705	4
7	0.6817	0.4648	0.4813	0.5426	8
8	0.5882	0.6767	0.3929	0.5526	7
9	0.8138	1.0000	0.6749	0.8296	1
10	0.4641	0.3333	0.3333	0.3769	18
11	0.4460	0.4298	0.4766	0.4508	14
12	0.4220	0.5410	0.3542	0.4391	15
13	0.4321	0.3965	0.4171	0.4152	17
14	0.3333	0.5410	0.4288	0.4344	16
15	0.5982	0.7296	0.4147	0.5808	6
16	0.6085	0.4648	0.3520	0.4751	13
17	0.6941	0.6767	0.6448	0.6719	3
18	0.4791	1.0000	1.0000	0.8264	2

TABLE 10.5
Response Table for the GRG

Symbol	End Milling Machining Parameter	L-1	L-2	L-3	Max–Min
A	Coolant	0.5821	0.5190	–	0.5821
B	Feed rate	0.4629	0.5391	0.6497	0.6497
C	Depth of cut	0.4717	0.5352	0.6447	0.6447
D	Speed	0.5922	0.5542	0.5052	0.5922

The feed rate is the most influential process parameter, as shown in Table 10.5 by the values of max–min, where GRG values of feed rate are more significant than any other process parameters. The remaining process parameters C, D, and A, as shown in Table 10.5, are listed in descending order of impact on multi-performance.

GRA selects the best end milling process settings. Each end milling parameter's maximum grey relational levels are A_1, B_3, C_3, and D_1, as shown in Figure 10.5. As an outcome, machining parameters in end milling with values $A_1B_3C_3D_1$ are a multi-performance-optimized combination.

10.4.1.1 Analysis of Variance (ANOVA)

ANOVA is incorporated to identify the most important factors impacting end milling. The statistical technique F-test was used to assess the impact of these input parameters on quality attributes. When the value exceeds four, adjustments in end milling parameters have a considerable effect on quality.

As shown in Table 10.6, feed is the most critical parameter, contributing 36.6708%, followed by depth of cut (31.8145%), speed (7.9137%), and coolant (6.2197%).

10.4.1.2 Validation Test

Compared to the predicted GRGs, there is a 1.52% increase in experimental grade, which is satisfactory, as shown in Table 10.7.

10.5 OPTICAL MICROGRAPHS

The ninth experimental work piece's microstructure test was finished since it is the most consistent and best approach to discovering microfaults, microcracks, and changes linked to microstructure. Many surface imperfections such as microfractures and microfaults depend on machining conditions. To inspect surface features, better metallographic techniques and optical microscope instruments are required. In addition to the data mentioned above, the microstructure was studied on SS-304 to determine the maximum surface strain. Figure 10.6a and b shows the microstructure images collected with an optical microscope (make: Rsamet Unitron) at 600×. The microstructure shown in Figure 10.6a displays pearls and polygonal ferrite particles. Figure 10.6b shows the microstructure of pearls and polygonal ferrite particles following end milling. Particles were 4–5 in number and tiny in size.

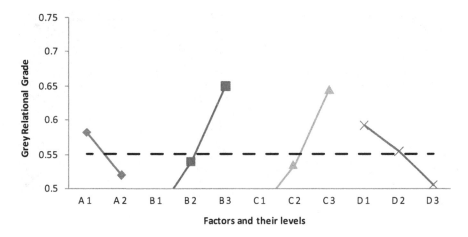

FIGURE 10.5 Main effects plot for GRG.

TABLE 10.6
ANOVA Grade Values

Symbol	Source	Sum of Squares	DoF	Mean Square	F-value	%C
A	Coolant	0.0180	1.0000	0.0180	3.5784	6.2197
B	Feed rate	0.1059	2.0000	0.0529	10.5489	36.6708
C	Depth of cut	0.0919	2.0000	0.0459	9.1519	31.8145
D	Speed	0.0229	2.0000	0.0114	2.2765	7.9137
Error		0.0502	10.0000	0.0050		17.3813
Total		0.2888	17.0000			100.0000

TABLE 10.7
Outcomes of Validation Test

	Optimum Machining Parameter		
	Prediction	Experiment	% Change
Level	$A_1B_3C_3D_1$	$A_1B_3C_3D_1$	
SR		0.8138	
MRR		1.0000	
MH		0.6749	
GRG	0.8171	0.8296	1.52

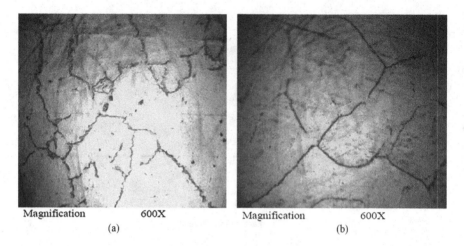

FIGURE 10.6   Microstructure photographs: (a) Before end milling; (b) after end milling.

10.6   CONCLUSIONS

The concluding remarks are as given below:

- The ideal setup of end milling process parameters is $A_1B_3C_3D_1$, which yields optimum multi-performance features with coolant on feed rate 2,500 mm/min, depth of cut 0.4 mm, and speed of machining 1,500 rpm.
- ANOVA determined that the contribution of feed rate, depth of cut, machining speed, and coolant is 36.67%, 31.81%, 7.91%, and 6.21%, respectively.
- This strategy yields a 1.52% improvement. As a result, Taguchi is a unique method for reducing costs and time.
- The structure of pearls and polygonal ferrite particles may be seen in the micrographs (before and after milling). There were a few rough areas with gritty particles as well. Particle counts ranged from 4 to 5, with small particles being the most prevalent.

In the future, fuzzy logic, genetic algorithm (GA), and numerous nature-inspired algorithms may be utilized for optimization.

REFERENCES

1. Zhang J. Z., Chen J. C., & Kirby E. D. (2007). Surface roughness optimization in an end-milling operation using the Taguchi design method. *Journal of Materials Processing Technology*, 184(1–3), 233–239.
2. Parmar J. G. & Makwana A. (2012). Prediction of surface roughness for end milling process using Artificial Neural Network. *International Journal of Modern Engineering Research (IJMER)*, 2(3), 1006–1013.
3. Jawahir I. S., Brinksmeier E., M'saoubi R., Aspinwall D. K., Outeiro J. C., Meyer D., Umbrello, D.,& Jayal A. D. (2011). Surface integrity in material removal processes: Recent advances. *CIRP Annals*, 60(2), 603–626.

4. Hamdan A., Sarhan A. A., & Hamdi M. (2012). An optimization method of the machining parameters in high-speed machining of stainless steel using coated carbide tool for best surface finish. *The International Journal of Advanced Manufacturing Technology*, 58(1), 81–91.

5. Prashar D., & Arora K. (2020). Design of two area load frequency control power system under unilateral contract with the help of conventional controller. *IJICTDC*, 5, 22–27.

6. Kumar G., Kumar M., & Tomer A. (2021). Optimization of end milling machining parameters of SS 304 by Taguchi technique. In Muzammil M., Chandra A., Kankar P.K., & Kumar H. (eds) *Recent Advances in Mechanical Engineering* (pp. 683–689). Springer, Singapore.

7. Kumar M., Kumar G., Singh O. P., & Tomer A. (2021). Multi performance optimization of parameters in deep drilling of SS-321 by Taguchi-based GRA. In *Recent Advances in Mechanical Engineering* (pp. 675–681). Springer, Singapore.

8. Karnwal A., Hasan M. M., Kumar N., Siddiquee A. N., & Khan Z. A. (2011). Multi-response optimization of diesel engine performance parameters using thumba bio-diesel-diesel blends by applying the Taguchi method and grey relational analysis. *International Journal of Automotive Technology*, 12(4), 599–610.

9. Patel K., Arora K., & Kaur P. "Power monitoring system in solar power plant using LabVIEW," *2019 2nd International Conference on Intelligent Computing, Instrumentation and Control Technologies (ICICICT)*, 2019, pp. 1011–1015, doi:10.1109/ICICICT46008.2019.8993249.

10. Ho C. Y., & Lin Z. C. (2003). Analysis and application of grey relation and ANOVA in chemical–mechanical polishing process parameters. *The International Journal of Advanced Manufacturing Technology*, 21(1), 10–14.

11. Singh O. P., Kumar G., & Kumar M. (2019). Multi Performance optimization of shoulder milling process parameters of AA6063 T6 aluminium alloy by Taguchi Based GRA. *International Journal of Innovative Technology and Exploring Engineering*, 8, 420–425.

12. Singh O. P., Kumar G., & Kumar M. (2019). Role of Taguchi and grey relational method in optimization of machining parameters of different materials: A review. *Acta Electronica Malaysia (AEM)*, 3(1), 19–22.

13. Siddiquee A. N., Khan Z. A., & Mallick Z. (2010). Grey relational analysis coupled with principal component analysis for optimisation design of the process parameters in in-feed centreless cylindrical grinding. *The International Journal of Advanced Manufacturing Technology*, 46(9), 983–992.

14. Khan Z. A., Kamaruddin S., & Siddiquee A. N. (2010). Feasibility study of use of recycled high-density polyethylene and multi response optimization of injection moulding parameters using combined grey relational and principal component analyses. *Materials & Design*, 31(6), 2925–2931.

15. Jangra K. K., Sharma N., Khanna R., & Matta D. (2016). An experimental investigation and optimization of friction stir welding process for AA6082 T6 (cryogenic treated and untreated) using an integrated approach of Taguchi, grey relational analysis and entropy method. *Proceedings of the Institution of Mechanical Engineers, Part L: Journal of Materials: Design and Applications*, 230(2), 454–469.

16. Sharma N., Khanna R., Sharma Y. K., & Gupta R. D. (2019). Multi-quality characteristics optimisation on WEDM for Ti-6Al-4V using Taguchi-grey relational theory. *International Journal of Machining and Machinability of Materials*, 21(1–2), 66–81.

11 Energy Storage Devices Based on 2D Phosphorene as an Electrode Material

A. Gomathi, T. Prabhuraj, P. Maadeswaran, and K. A. Ramesh Kumar
Periyar University

CONTENTS

DOI: 10.1201/9781003242277-11

11.1 INTRODUCTION

For nearly a century, scientists have known about black phosphorus (BP). In recent years, intellectuals from all over the world have taken notice of phosphorene's potential to exist in a two-dimensional (2D) form. Phosphorene is expected to get the most focus in the post-graphene period due to its extraordinary and unique band structures; carrier transport; and optical, thermal, and mechanical performance. Although environmental advances have extensively been spread throughout human civilization in recent decades, their performance does not meet the criteria for future growth. The potential is used in electrochemical energy storage, such as Li/Na. The achievements and problems of K-/Mg-/Na-ion and Li–S batteries are methodically summarized. A diverse array of S domains, including ultrafast laser biophotonics, energy storage devices, optoelectronic solar cells, and nanosensors, have extensively been investigated using two-dimensional black phosphorene [1,2]. Due to its vast surface area, superior electric conductivity, and ideally high theoretical value, 2D BP has greatly been explored as an electrode material and has significantly enhanced energy storage device performance.

Few-layer phosphorene has inherently tunable direct band gaps (ranging from 0.3 to 2 eV), high carrier mobility (200–1,000 cm²/V s), great mechanical flexibility, and anisotropic electronic and phonon dispersion. As a result of its intriguing properties, few-layer phosphorene is used in a broad array of applications, including field-effect transistors [3], photodetectors [4], photovoltaic devices [5,6], biosensors, gas sensors [7], superconductors, thermoelectric conversion [8], nano-electromechanical resonators, photosensitizers, photocatalysts, and electrocatalysts [9].

Phosphorene, a two-dimensional layered substance made up of layers of black phosphorus, was revived as an energy storage material. Black phosphorus, better known as phosphorene, is a monolayer material [10].

11.2 FUNDAMENTAL PROPERTIES OF PHOSPHORENE

11.2.1 Band Structures

Black phosphorous is a common direct-band-gap semiconductor. When thinned to few-layer and monolayer thicknesses, it retains a direct band gap, as shown in Figure 11.1a. The band gap of phosphorene is sensitive to the number of layers, as shown in Figure 11.1b [11]. The electron–electron interaction dominates semiconductor basic band gaps, which may be accurately described using HSE06 or the GW technique [11].

11.2.2 Carrier Transport

Two-dimensional semiconductors belonging to group-VA due to their superior carrier transport characteristics have led to their possible use in electrical and optoelectronic devices. It is universally acknowledged that mobility calculations are used to assess carrier transport properties [11].

11.2.3 Optical Properties of Phosphorene

Two-dimensional group-VA semiconductors exhibit unique optical characteristics. Because of this, the anticipated absorption spectrum is anisotropic. In the armchair,

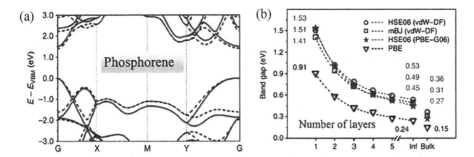

FIGURE 11.1 (a) Band structures of black phosphorene monolayer calculated with the HSE06 function (black solid lines) and the mBJ potential (black dashed lines), respectively Functional use for structural optimization is shown. The electron–electron interaction dominates the fundamental band gap of semiconductors, which can be adequately characterized by HSE06 or the GW. (b) Progression of the direct-band-gap function as a sample of the thickness method. (Reproduced with permission from Qiao, J., et al., *Nat. Commun.*, 5, 1–7, 2014 [11].)

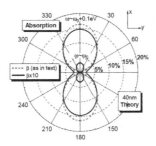

FIGURE 11.2 Polar representation of the absorption coefficient $A(x)$ for a 40-nm intrinsic black phosphorus film for normal incident light with excitation energies at the band gap ω0, and larger. α is the light polarization angle. $A(x)$ is plotted for two values of interband coupling strengths. (Reproduced with permission from Low, T., et al., *Phys. Rev. B: Condens. Matter Mater. Phys.*, 90, 075434, 2014 [13].)

the light is linearly polarized and in zigzag directions. When it comes to absorption in ultrathin layers, the band edge falls off quickly in the armchair direction, indicating that the band gap is found in the armchair direction. However, the highest monolayer potential is located at 3.14 eV and only slightly decreases in thickness. As optical detection of crystallographic orientation and optical activation of anisotropic transport characteristics are both feasible with an in-plane linear diachronic, a variety of uses are possible [12]. The phosphorene monolayer is transparent to light in the same energy range in the zigzag direction, which includes the infrared and a portion of the visible-light regime [13] (Figure 11.2).

11.2.4 THERMAL PROPERTIES

These materials have been researched using both theoretical and experimental techniques to determine their thermal properties [14–17]. Phosphorene's linear thermal

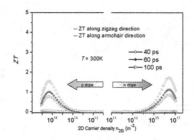

FIGURE 11.3 The thermoelectric figure of merit according to the doping density at $T = 300$ K. Different long-wave relaxation times of phonons are included. The unit of the doping density is m⁻². (Reproduced with permission from Fei, R., et al., *Nano Lett.*, 14, 6393–6399, 2014 [19].)

expansion coefficients are expected to be extremely anisotropic in both the zigzag and armchair directions (up to 20%) [16]. There exist phonon relaxation periods that depend on the direction of group velocities in black phosphorene [18]. A chair's electrical conductivity is greater than a zigzag's [19] (Figure 11.3).

11.2.5 MECHANICAL PROPERTIES

Two-dimensional semiconductors belonging to group-VA materials exhibit unique mechanical characteristics due to their puckered and buckled structures [20–22]. Wei and Peng [21] carried out first-principles calculations to discover that the zigzag direction has the highest Young's modulus of 166 GPa, while the armchair direction has the lowest at 44 GPa (Figure 11.4). Black phosphorus may be stretched by up to 30% when seen from an armchair. Jiang and Park [20] concluded that the puckered structure of phosphorene gave rise to a negative Poisson's ratio in the out-of-plane vector. Due to this remarkable feature, phosphorene promises as an auxetic material.

11.3 TUNABLE ELECTRONIC PROPERTIES

Material characteristics may be tuned in two-dimensional semiconductors belonging to group-VA nanosheets, which opens various applications. To efficiently modify their essential properties, many methodologies including strain, electric field, doping, defect, chemical fictionalization, and heterostructure have theoretically been proposed.

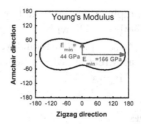

FIGURE 11.4 Direction dependence of Young's modulus of phosphorene. (Reproduced with permission from Wei, Q. and Peng, X. *Appl. Phys. Lett.*, 104, 251915, 2014 [21].)

11.3.1 STRAIN/ELECTRIC FIELD

Ultrathin 2D materials have an immense stretching ability and can reversibly endure significant structural deformation, as is extensively known. Strain has been proposed as a viable way for modulating the electrical characteristics of 2D materials, with the potential to broaden their applications [23–25]. Under external strain, the band gap of phosphorene was investigated [25]. The band gap value drops with an increase in strain. Interestingly, the band gap value increases to a maximum at critical tensile strain (4%) and subsequently decreases with increasing strain in the zigzag direction. The dramatic shift under external tension is also due to the effective mass of phosphorus. The imposed biaxial strain, according to Fei and Yang, can modify the preferred direction of electron transport [26]. In the presence of an external electric field, Dai and Zeng investigated the few-layer phosphorene's band gap [6]. The band gap reduces (from 0.78 to 0.56 eV), with enhanced electric field, whereas monolayer phosphorene hardly changes somewhat (from 0.86 to 0.78 eV).

11.3.2 DEFECT

Defects can adversely impact the electrical characteristics of two-dimensional semiconductors belonging to group-VA materials and induce magnetism in nonmagnetic pure phosphorene, according to theoretical simulations [27, 28]. The electronic band structures and bias of certain defective phosphorene systems are reliant on their transport properties. Umar Farooq et al. [29] investigated the electronic band topologies and bias-dependent transport characteristics of different phosphorene systems with impurities. Hu et al. [30] explored 10 different kinds of phosphorene point defects, such as Stone–Wales flaws, single vacancies, and double vacancies in phosphorene.

11.3.3 SURFACE FUNCTIONALIZATION

Chemical and physical adsorptions on the surface of two-dimensional semiconductors belonging to group-VA materials can be used to enhance their intrinsic properties. As a result of external adsorption doping, the electronic properties of phosphorene can turn the semiconducting phosphorene into a metal or a half-metal, as well as the spin-polarized semiconductor into a non-polarized semiconductor by selective doping. B, C, N, F, and Li adsorption on phosphorene leads to a Fermi level shift and metallic electronic structures [31–33]. A 1B magnetic moment results from the absorption of NO and NO_2. Furthermore, NH_3 (NO) adsorption on phosphorene might change the resistance and cause the current to increase (reduce) because of transport calculations. Phosphorene oxides were explored physically and chemically by Wang et al. [34]. Phosphorene oxides are stable at ambient circumstances, according to their phonon spectrum. They discovered that phosphorene's band gap is influenced by its degree of O-functionalization. External strain or an electric field can also be used to fine-tune the band gap of phosphorene oxide. Black phosphorous has a slender band gap semiconductor that can undergo a semiconductor-to-semiconductor quantum phase transition under strain, according to first-principles calculations. They also discovered exotic phenomena by developing effective models for the phase transition and unique emergent fermions. BPO will help researchers better understand quantum phase transitions and fermions' fundamental features, and it can be applied in the field of nanotechnology [35].

11.3.4 HETEROSTRUCTURES

Van der Waals heterostructures (vdWHs) have been the subject of numerous investigations. Graphene/BN, graphene/TMDCs, graphene/phosphorene, BN/TMDCs, BN/phosphorene, TMDCs/phosphorene, and TMDCs/MXenes are examples of vdWHs with two distinct monolayers that are free of dangling bonds. Various investigations have shown that these vdWHs have many innovative optical and electrical features that are applied to create perfect optoelectronic and electronic devices. Phosphorene [36, 37] provides prospective building blocks for vdWHs as a unique two-dimensional semiconductor belonging to group-VA family. Guo et al. studied the hetero-bilayers of phosphorene with a TMDC monolayer, which has semiconductor characteristics, and discovered that their band gaps can be decreased by a vertical electric field in the phosphorene-based heterostructure [38]. The power conversion efficiency of the phosphorene/MoS_2 hetero-bilayer was up to 17.5%. vdWHs have two advantages in terms of computing outcomes. A graphene/phosphorene/graphene heterostructure was also studied by Wang et al. [39]. (i) The thermionic transport barriers can be adjusted by modifying the number of layers and (ii) the thermal conductivity across these non-covalent structures is low. Layered van der Waals structures are commonly thought to have the potential to be employed in solid-state energy conversion devices. Yu et al. then studied the features of a van der Waals phosphorene p–n heterostructure and discovered that it had an indirect band gap and a type II band alignment.

11.4 SYNTHESIS OF PHOSPHORENE

BP can be synthesized in bulk using a variety of methods, including high-pressure synthesis, recrystallization from bismuth (Bi) flux, chemical vapor transport, and mechanical milling.

11.4.1 HIGH-PRESSURE ROUTE

Bridgman [40] described the discovery of black phosphorus as the result of regular white phosphorus (white P) being pushed under high hydrostatic pressure to convert to red phosphorus (red P). The phenomenal change of white to black phosphorus occurred when pressure (11,000–13,000 kg/cm^2) was applied to white phosphorus at room temperature through a kerosene medium at 200°C in an oil bath regulated by a thermostat. Bulk BP was also made red by heating phosphorus to 1,000°C and slowly cooling it to 600°C at a cooling rate of 100°C/h [40] under a constant pressure of 10 kbar. Synthetic BP should be kept in a completely inert environment. The synthesis of BP was reported using a high-temperature, high-pressure (HTHP) process in a sintered boron nitride chamber employing a cubic anvil high-pressure equipment operating at a pressure of 2–5 GPa, forming cylindrical capsules from white and red phosphorus powder blocks (3 mm thick and 10 mm in diameter) [41].

11.4.2 Recrystallization from Bismuth Flux

In 1965, Brown and Rundqvist [42] first published the bismuth flux method for producing needle-shaped black phosphorene monocrystals from a solution of white phosphorene in liquid bismuth. According to Iwasaki et al. [43], the needle-shaped black phosphorene crystals created by the Biflux process exhibited a 2D Anderson localization of electrical properties at low temperatures. Baba et al. [44] used an enhanced Biflux technique to generate black phosphorene crystals in various forms from a solution of white phosphorus in liquid Bi, with lesser chemical contaminants. Highly pure white phosphorene is also difficult to obtain because of its chemical activity [44]. In a quartz ampoule, the white phosphorene (melting point 44.1°C) was melted at 80°C before being introduced to the Bi (heated to 300°C). After that, the ampoule was heated in an electric furnace to 400°C for 48 hours before being cooled.

11.4.3 Chemical Vapor Transport

The procedure of creating single crystals of BP using chemical vapor transfer (CVT) [45] is described in detail below. The charged end of the ampoule was placed horizontally in the center of a single-zone tube furnace. The ampoule was slowly heated to 873 K for about 10 hours and then held at that temperature for another 24 hours. Subsequently, the temperature was reduced to 773 K at a rate of 40 K/h. At the cool end of the ampoule, the BP single crystals were synthesized and further solidified in the shape of flakes. Due to its good yield and safety under nontoxic circumstances, the low-pressure technique with the use of a mineralizer was used as the reaction promoter proved successful [45].

Mineralization converts red phosphorus to BP, which produces red phosphorus and Au as by-products [46]. Nilges et al. [47] used the mineralizer SnI4 synthesized from tin powder and iodine in 25 mL toluene (starting materials) and refluxed it for 30 minutes; the synthesis technique they used for the low-pressure method is shown below. Au and Sn were synthesized from an equimolar combination of gold and tin in a sealed evacuated silica ampoule, and Au and Sn were used as a binary precursor to accelerate the production of polyphosphide Au3SnP7 at elevated temperatures prior to the transport process. The initial components were melted with an H_2/O_2 burner prior to the growth procedure. Red phosphorus, Au Sn, and SnI4 were packed in a silica ampoule (10 cm long; 10 mm in diameter), which were then evacuated to a pressure of 10–3 mbar and heated in a muffle furnace (873 K, 23 hours), yielding BP crystals (>1 cm). The Sn-to-SnI4 ratio is the most important factor in the formation of high-quality BP bulk crystals. To eliminate mineralizer residues, the final BP product was collected and washed repeatedly with hot toluene and acetone [48]. A modified mineralizer-aided short-way transport reaction using red phosphorus, Sn/SnI4 as the mineralization additive to promote short reaction durations, and high-quality big BP crystals was also reported in the literature [49].

11.4.4 Mechanical Milling

BP can be manufactured using a mechanical milling process involving a mixer mill and a planetary ball mill device using red phosphorus powder as the raw material

[50]. The procedure was carried out for one hour in an Ar environment in a stainless steel jar with ten stainless steel balls (10 mm or 12.7 mm in diameter). A similar milling procedure (a mixer mill) was utilized to fabricate composites from BP and acetylene black (AB) (80 wt.% BP and 20 wt.% AB). The mixer mill equipment generated greater crystallinity black phosphorus than the other two ball mill apparatuses [51]. Red phosphorus was washed with distilled water and a 5% sodium hydroxide solution to remove oxides. A stainless steel jar containing red phosphorus and various-sized stainless steel balls was sealed and rotated at 400 rpm for 12 hours in an Ar-filled glove box.

11.5 ENERGY STORAGE DEVICES—PHOSPHORENE

11.5.1 Li-ion Batteries

In early 2007, the electrochemical characteristics of BP as an anode for Li-ion batteries were announced. It has a large charge capacity of 1,279 mAh/g and a coulombic efficiency of 57% [52]. Because of its huge specific surface area relative to black phosphorene, few-layer phosphorene may have high initial irreversible capacity owing to electrolyte breakdown and the development of a solid electrolyte interface (SEI) film on the phosphorene surface. The electrochemical activity of BP may improve due to relevant degree of crystallite and small size [53]. The BP anode in the $Li_2S–P_2S_5$ solid electrolyte had a better coulombic efficiency in a liquid electrolyte [50].

The electrochemical activity of the BP anode was used by the researchers to demonstrate that phosphorene is a potential anode material for Li-ion batteries. When studying the binding and diffusion characteristics of Li-ions on the phosphorene surface, density functional theory (DFT) simulations were first employed [54,55]. It was found that strong binding was formed with phosphorene atoms and Li atoms with a formation energy of approximately −1.9 eV (Figure 11.5a). A high charge shortage at Li atoms and a charge excess nearby P atom, implying a significant electron transfer from Li to phosphorene [55]. Due to a significant electron transfer from Li to phosphorene, there was a significant charge shortage of Li atoms and an excess charge surrounding nearby P atoms [55]. The single Li atom was located above the three P atoms in the triangle's center. The bonding distances between Li and the three closest P atoms were 2.54, 2.54, and 2.45 A, respectively [56]. Li was in a cationic state because its 2s1 electron completely transferred to phosphorene, thereby shifting the Fermi energy level back to the conduction band. The transition from semiconductor to conductor is critical for Li-ion battery high-rate anodes. Phosphorene doped with Si atoms enhanced the binding energy with Li atoms. With a binding energy of 2.40 eV (Figure 11.5b) [57], the Li atom atop the triangle of Si and two P atoms produced the greatest stable binding energy. It was substantially lower than the virgin phosphorene value of 1.90 eV [55]. The greater the negative value of binding energy, the more energetically exothermic the process became. The applications of phosphorene/graphene heterostructures as anode materials in Li-ion batteries have various applications. During the litigation process, Li atoms were discovered to be disposed to insert the interlayer of the phosphorene/graphene heterostructure with a binding energy of 2.59 eV (Figure 11.5c) rather than adsorption on the outer surface of phosphorene [57] (Figure 11.5d).

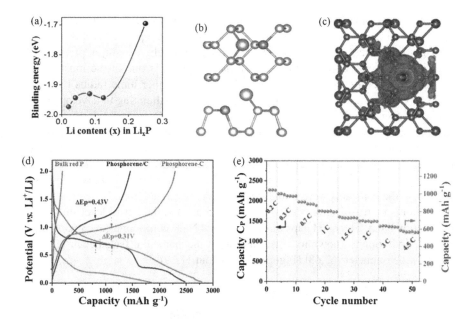

FIGURE 11.5 (a) Binding energies with increasing Li content on phosphorene. (Reproduced with permission from Li, W., et al., *Nano Lett.* 15, 1691–1697, 2015 [55]. Copyright 2015, American Chemical Society.) (b) Optimized configurations of one Li adsorbed on Si-doped phosphorene. (Reproduced with permission from Zhao, S., et al., *J. Mater. Chem. A* 2, 19046–19052, 2014 [56]. Copyright 2014, The Royal Society of Chemistry.) (c) Charge density difference of Li atoms incorporation into interlayer of graphene/phosphorene. (Reproduced with permission from Guo, G.C., et al., *J. Phys. Chem. Lett.*, 6, 5002–5008, 2015 [57]. Copyright 2015, American Chemical Society.) (d) The charge/discharge profiles of bulk red P, phosphorene/C mixture, and phosphorene–C composite at the initial cycle. (Reproduced with permission from Sun, J., et al., *Nano Lett.*, 14, 4573–4580, 2014 [58]. Copyright 2014, American Chemical Society.) (e) The rate capability of the phosphorene–C composite at various current densities. (Reproduced with permission from Sun, J., et al., *Nano Lett.*, 14, 4573–4580, 2014 [58]. Copyright 2014, American Chemical Society.)

The tight bond between phosphorene/graphene and the Li atom may prevent the production of metallic Li dendrites during the litigation process. It will also increase the safety of the Li-ion battery under study. At various current densities, the phosphorene–carbon composite demonstrated higher charge/discharge capacities than the physical mixture. The P–C bonds also increased coulombic efficiency and decreased potential hysteresis (E) of the discharge and charge plateaus. Even when the current rate increased to 4.5 C, the composite base of the chemical mixture was still retained at a specific capacity of 1,240 mAh/g (Figure 11.5e), which strongly demonstrated the advantage of forming a special chemical bond in phosphorene-based composites.

11.5.2 Na-ion Batteries

The non-aqueous Na-ion battery is the most important of all materials. During charge/discharge operations, Na-ions rock back and forth between the anode and cathode, like Li-ion batteries [59, 60]. Nonetheless, because Na$^+$ ion has a greater ionic radius (1.02 A) than Li$^+$ ion (0.76 A), developing acceptable Na-host materials with optimum electrochemical characteristics is difficult. Few-layer phosphorene, with a greater interlayer spacing (5.4 A) than graphite, is one of the most promising anode materials for Na-ion storage (3.4 A). The higher ionic radius of Na$^+$ ion (1.02 A) compared to Li$^+$ ion (0.76 A) makes it more challenging to develop suitable Na-host materials with electrochemical characteristics. On monolayer phosphorene, Na diffusion was rapid and anisotropic, with a less energy barrier of only 0.04 eV [61]. The total density of states (DOS) for various Na–phosphorene combinations revealed a semiconductor metal transition as the concentration of Na on phosphorene rises as shown in Figure 11.6a. When the concentration of Na was low (such as NaP$_{16}$), the Na states were found in the conduction band area. When the number of Na atoms increases, more charge transfer occurs and the Fermi level shifts to the conduction band. The band gaps of the NaP$_2$ and NaP$_4$ compounds eventually disappeared. Sodiation is known to cause significant mechanical effects because Na-ions have a wide diameter [62,63] (Figures 11.6b,c and 11.6d).

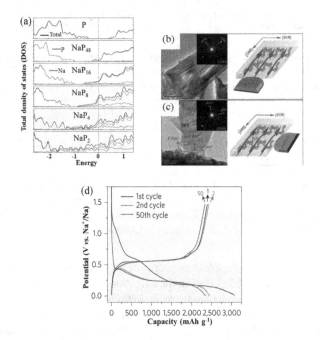

FIGURE 11.6 (a) The total density of states (DOS) for different Na–phosphorene compounds. (Reproduced with permission from Kulish, V.V., et al., *Phys. Chem. Chem. Phys.*, 17, 13921–13928, 2015 [61]. Copyright 2015, The Royal Society of Chemistry.) (b) and (c) Na-ion transport in few-layer phosphorene with the contact interface normal to the [1 0 0] direction vs. the sodium source and parallel to the [1 0 0] direction vs the Na anode. (Reproduced with permission from Nie, A., et al., *Nano Lett.*, 16, 2240–2247, 2016 [62]. Copyright 2016, American Chemical Society.) (d) Galvanostatic discharge/charge curves of the phosphorene/graphene anode for the initial, second, and 50th cycles. (Reproduced with permission from Sun, J., et al., *Nat. Nanotechnol.*, 10, 980–985, 2015 [63]. Copyright 2015, Macmillan Publishers Limited.)

11.5.3 K-ion Batteries

K-ion batteries are employed in ion batteries because K has a comparable abundance as Na (23,000 ppm), while Li has 20 ppm in the Earth's crust. The redox potential of the K metal anode is lower than that of the Na metal anode. There are less costly alternatives to Li-ion batteries, such as K-ion batteries, which have redox potentials that are much closer to Li's (3.040 V vs. SHE; $K/K^+ = 2.936$ V vs. SHE) [64–67]. There are few materials that can store K-ions, due to their high ionic radius, which results in the battery's unique thermodynamics and dynamics [68,69]. In addition, it is noted that the radius of the K cation in solution. Because of the lower Lewis acidity of K-ion, anode materials used to store K-ions are smaller than those used to store Li and Na in solution. If you use liquid electrolytes, the Stokes radius of K-ions is lower than the Stokes radius of Na- and Li-ions. As a result, the ionic conductivities of solvated K-ion electrolyte are greater than those of Li- and Na-ions, suggesting faster diffusion kinetics at the electrolyte/electrode interface is possible. In K-ion batteries, the current collector for anodes can be as cheap as in Na-ion batteries. The evolution of anode materials and, to a large extent, how to design a high-performance K-ion full are both critical to avoid K dendrite formation. The final alloying phase for phosphorene-based anodes has recently been reported to be KP alloy with a theoretical capacity of 843 mAh/g. The associated phosphorene–carbon composites had a capacity of more than 600 mAh/g, which outperformed all or any state-of-the-art K-ion battery anodes [70]. The phosphorene–carbon anode's redox potential was at 0.6–0.7 V vs. K/K^+, which inhibited the formation of K dendrites while remaining low enough to form a high-voltage K-ion complete battery. In order to improve cycle stability, few-layered phosphorene–carbon composite was chemically bonded as mentioned for K-ion storage. The as-prepared phosphorene–carbon anode showed an outstanding cycle stability of up to 500 cycles. The carbon matrix was tightly anchored by the few-layer phosphorene. The composite's flexible carbon matrix not only buffers the severe volume shift, but also improves phosphorene's electronic conductivity. More crucially, the P–C and P–O–C bonds in the composite result in an intrinsic electrical connection and a well-knit structure, which considerably increases the structure's durability [71].

11.5.4 Li–S Batteries

The sulfur electrode possesses a high theoretical capacity (1,672 mAh/g). Li–S battery is considered as the promising material for its high energy density and low cost. It will become an important energy storage device after Li battery [72–75]. Severe capacity deterioration is unavoidable because of the significant volume growth and shuttle effect of soluble polysulfide. Lithium polysulfides (Li_2S_x, $x = 1$–6) were found to have binding energies of 2.49–0.92 eV on phosphorene surfaces (Figure 11.7a), which were substantially greater than the same binding energies on a carbon hexatomic ring network based on DFT calculations using van der Waals force corrections. After 500 cycles, the phosphorene/CNF/polysulfides electrode's specific capacity remained over 660 mAh/g, with only 0.053% average capacity degradation per cycle. This was a significant improvement over the CNF/polysulfides electrode without

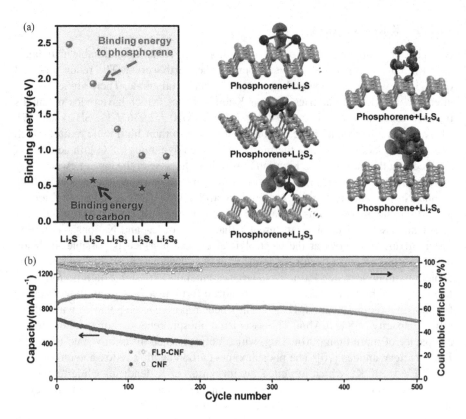

FIGURE 11.7 (a) DFT-calculated binding energy between lithium polysulfides and phosphorene and the corresponding atom positions and charge density. Red, green, and violet spheres represent sulfur, lithium, and phosphorus atoms, respectively. (Reproduced with permission from Zhu, L., et al., *Nano Lett.*, 16, 6548–6554, 2016 [35]. Copyright 2016, John Wiley & Sons, Inc.) (b) The cycling stability and coulombic efficiency. (For interpretation of the references to color in this figure legend, the reader is referred to the web version of the article.) (Reproduced with permission from Zhu, L., et al., *Nano Lett.*, 16, 6548–6554, 2016 [35]. Copyright 2016, John Wiley & Sons, Inc.)

phosphorene (Figure 11.7b). Furthermore, few-layer phosphorene has an electrical conductivity of 450 s/m, which is higher than metal oxides and carbon nitride-based Li–S battery substrates [76].

11.5.5 Mg-ion Batteries

DFT calculations were used to assess the viability of phosphorene as an electrode material for Mg-ion storage in terms of adsorption and diffusion of Mg atoms on its surface. First, two possible Mg-ion diffusion paths, along armchair or zigzag directions, were considered on the surface of phosphorene. On phosphorene, magnesium diffusion mobility in the zigzag direction is considerably quicker than in the armchair direction [77].

11.6 CONCLUSIONS

To summarize, few-layer phosphorene has been explored as promising electrode materials for various energy storage devices. For the application in the field of Li- and Na-ion batteries, phosphorene-based nanocomposite materials have shown high specific capacities. In addition, few-layer phosphorene can effectively adsorb and trap polysulfides, thus enhancing the electrochemical performance of Li-S battery. The polysulfide binding energies on the phosphorene surface range from 1 to 2.5 eV. Despite these properties, research on phosphorene-based energy storage devices is still in its early stage of development. Many energy storage systems need the bulk manufacturing of few layers of phosphorene. From an overall viewpoint, few-layer phosphorene is an emerging, but intriguing 2D material for energy storage devices. The phosphorene-based energy storage devices may get rapid development and achieve early practical applications.

REFERENCES

1. Wu, W.Z., Zhou, Y.J., Wang, J., Shao, Y.B., Kong, D. C, Gao, Y.C. and Wang, Y.G. (2020). The pump fluence and wave length dependent ultrafast carrier dynamics and optical non-linear absorption in black phosphorus Nano sheets, *Nano Photonics* 9, 2033–2043.
2. Guo, W.L., Dong, Z., Xu, Y.J., Liu, C.L. and Wei D.C. (2020). Terahertz detection and imaging driven by the photo thermo electric effect in ultra-short-channel black phosphorus devices, *Adv. Sci.* 7, 1902699.
3. Li, L., Yu, Y., Ye, G.J., Ge, Q., Ou, X., Wu, H., Feng, D., Chen, X.H. and Zhang, Y. (2014). Lack phosphorus field-effect transistors, *Nat. Nano Technol.* 9, 372–377.
4. Youngblood, N., Chen, C., Koester, S.J. and Li, M. (2015). Waveguide-integrated black phosphorus photo detector with high responsivity and low dark current, *Nat. Photonics* 9, 247–252.
5. Jia, J., Xu, J., Park, B.H., Lee, E., Wang, H. and Lee, S. (2017). Multifunctional homogeneous lateral black phosphorus junction devices, *Chem. Mater.* 29, 3143–3151.
6. Dai, J. and Zeng, X.C. (2014). Bilayer phosphorene: Effect of stacking order on band gap and its potential applications in thin-film solar cells, *J. Phys. Chem. Lett.* 5, 1289–1293.
7. Cui, S., Pu, H., Wells, S.A., Wen, Z., Mao, S., Chang, J., Hersam, M.C. and Chen, J. (2015). Ultra high sensitivity and layer-dependent sensing performance of phosphorene-based gas sensors, *Nat. Commun.* 6, 8632.
8. Qin, G.Z., Yan, Q.B., Qin, Z.Z., Yue, S.Y., Cui, H.J., Zheng, Q.R. and Su, G. (2014) Hinge-like structure induced unusual properties of black phosphorus and new strategies to improve the thermoelectric performance, *Sci. Rep.* 4, 6946.
9. Wang, Z., Jia, H., Zheng, H., Yang, R., Wang, Z., Ye, G.J., Chen, X.N., Shan, J. and Feng, P.X.L. (2015). Black phosphorus Nano electromechanical resonators vibrating at very high frequencies, *Nano Scale* 7, 877–884.
10. Pang, J., Bachmatiuk, A., Yin, Y., Trzebicka, B., Zhao, L., Fu, L., Mendes, R.G., Gemming, T., Liu, Z. and Rummeli, M.H. (2018). Applications of phosphorene and black phosphorus in energy conversion and storage devices, *Adv. Energy Mater.* 8, 1702093.
11. Qiao, J., Kong, X., Hu, Z.X., Yang, F. and Ji, W. (2014). High-mobility transport anisotropy and linear dichroism in few-layer black phosphorus, *Nat. Commun.* 5, 1–7.
12. Tran, V., Soklaski, R., Liang, Y. and Yang, L. (2014). Layer-controlled band gap and anisotropic excitons in few-layer black phosphorus, *Phys. Rev. B: Condens. Matter Mater. Phys.* 89, 235319.

13. Low, T., Rodin, A.S., Carvalho, A., Jiang, Y., Wang, H., Xia, F. and Castro Neto, A.H. (2014). Tunable optical properties of multilayer black phosphorus thin films, *Phys. Rev. B: Condens. Matter Mater. Phys.* 90, 075434.

14. Zeraati, M., Vaez Allaei, S.M., Abdolhosseini Sarsari, I., Pourfath, M. and Donadio, D. (2016). Highly anisotropic thermal conductivity of arsenene: An ab initio study, *Phys. Rev. B: Condens. Matter Mater. Phys.* 93, 085424.

15. Cheng, L., Liu, H., Tan, X., Zhang, J., Wei, J., Lv, H., Shi, J. and Tang, X. (2014). Thermoelectric properties of a monolayer bismuth, *J. Phys. Chem. C* 118, 904–910.

16. Aierken, Y., Çakır, D., Sevik, C. and Peeters, F.M. (2015). Significant effect of stacking on the electronic and optical properties of few-layer black phosphorus, *Phys. Rev. B: Condens. Matter Mater. Phys.* 92, 081408.

17 Cheng, L., Liu, H.J., Zhang, J., Wei, J., Liang, J.H., Jiang, P.H., Fan, D.D., Sun, L. and Shi, J. (2016). High thermoelectric performance of the distorted bismuth (110) layer, *Phys. Chem. Chem. Phys.* 18, 17373–17379.

18. Zhu, L., Zhang, G. and Li, B. (2014). Coexistence of size-dependent and size-independent thermal conductivities in phosphorene, *Phys. Rev. B: Condens. Matter Mater. Phys.* 90, 214302.

19. Fei, R., Faghaninia, A., Soklaski, R., Yan, J.A., Lo, C. and Yang, L. (2014). Enhanced thermoelectric efficiency via orthogonal electrical and thermal conductances in phosphorene, *Nano Lett.* 14, 6393–6399.

20. Jiang, J.W. and Park, H.S. (2014). Negative Poisson's ratio in single-layer black phosphorus, *Nat. Commun.* 5, 4727.

21. Wei, Q. and Peng, X. (2014). Superior mechanical flexibility of phosphorene and few-layer black phosphorus, *Appl. Phys. Lett.* 104, 251915.

22. Wang, L., Kutana, A., Zou, X. and Yakobson, B.I. (2015). Electro-mechanical anisotropy of phosphorene, *Nanoscale* 7, 9746–9751.

23. Cai, Y., Ke, Q., Zhang, G., Feng, Y.P., Shenoy, V.B. and Zhang, Y.W. (2015). Giant phononic anisotropy and unusual anharmonicity of phosphorene: Interlayer coupling and strain engineering, *Adv. Funct. Mater.* 25, 2230–2236.

24. Rodin, A.S., Carvalho, A. and Castro Neto, A.H. (2014). Strain-induced gap modification in black phosphorus, *Phys. Rev. Lett.* 112, 176801.

26. Fei, R. and Yang, L. (2014). Strain-engineering the anisotropic electrical conductance of few-layer black phosphorus, *Nano Lett.* 14, 2884–2889.

27. Liu, Y., Xu, F., Zhang, Z., Penev, E.S. and Yakobson, B.I. (2014). Two-dimensional mono-elemental semiconductor with electronically inactive defects: The case of phosphorus, *Nano Lett.* 14, 6782–6786.

28. Yang, L., Song, Y., Mi, W. and Wang, X. (2016). The electronic structure and spin–orbit-induced spin splitting in antimonene with vacancy defects. *RSC Adv.* 6, 66140–66146.

29. Umar Farooq, M., Hashmi, A. and Hong, J (2015). Anisotropic bias dependent transport property of defective phosphorene layer, *Sci. Rep.* 5, 12482.

30. Hu, W. and Yang, J. (2015). Defects in phosphorene, *J. Phys. Chem. C* 119, 20474–20480.

31. Wang, G., Pandey, R and Karna, S.P. (2015). Effects of extrinsic point defects in phosphorene: B, C, N, O, and F adatoms, *Appl. Phys. Lett.* 106, 173104.

32. Hu, T. and Hong, J. (2015). First-principles study of metal adatom adsorption on black phosphorene, *J. Phys. Chem. C* 119, 8199–8207.

33. Kulish, V.V., Malyi, O.I., Persson, C. and Wu, P. (2015). Adsorption of metal adatoms on single-layer phosphorene, *Phys. Chem. Chem. Phys.* 17, 992–1000.

34. Wang, G., Pandey, R. and Karna, S.P. (2015). Phosphorene oxide: Stability and electronic properties of a novel two-dimensional material, *Nanoscale* 7, 524–531.

35. Zhu, L., Wang, S.S., Guan, S., Liu, Y., Zhang, T., Chen, G. and Yang, S.A. (2016). Blue phosphorene oxide: Strain-tunable quantum phase transitions and novel 2D emergent fermions, *Nano Lett.* 16, 6548–6554.

36. Avsar, A., Vera-Marun, I.J., Tan, J.Y., Watanabe, K., Taniguchi, T., Castro Neto, A.H. and Ozyilmaz, B. (2015). Air-stable transport in graphene-contacted, fully encapsulated ultrathin black phosphorus-based field-effect transistors. *ACS Nano* 9, 4138–4145.
37. Padilha, J.E., Fazzio, A. and Silva, A.J. (2015). van der Waals heterostructure of phosphorene and graphene: Tuning the Schottky barrier and doping by electrostatic gating, *Phys. Rev. Lett.* 114, 066803.
38. Guo, H., Lu, N., Dai, J., Wuand, X. and Zeng, X.C. (2014). Tuning electronic and magnetic properties of early transition-metal dichalcogenides via tensile strain, *J. Phys. Chem. C* 118, 14051–14059.
39. Wang, X., Zebarjadi, M. and Esfarjani, K. (2016). First principles calculations of solid-state thermionic transport in layered van der Waals heterostructures, *Nanoscale* 8, 14695–14704.
40. Bridgman, P. (1914). Two new modifications of phosphorus, *J. Am. Chem. Soc.* 36, 1344.
41. Zhang, X., Xie, H., Liu, Z., Tan, C., Luo, Z., Li, H., Lin, J., Sun, L., Chen, W. and Xu, Z. (2015). Black phosphorus quantum dots, *Angew. Chem. Int. Ed.* 54, 3653.
42. Brown, A. and Rundqvist, S. (1965). Refinement of the crystal structure of black phosphorus, *Acta Crystallogr.* 19, 684.
43. Iwasaki, N., Maruyama, Y., Kurihara, S., Shirotani, I. and Kinoshita, M. (1985). Negative magnetoresistance and Anderson localization in black phosphorus single crystal, *Chem. Lett.* 1, 119.
44. Baba, M., Izumida, F., Takeda, Y. and Morita, A. (1989). Preparation of black phosphorus single crystals by a completely closed bismuth-flux method and their crystal morphology, *J. Appl. Phys.* 28, 1019.
45. Ge, S., Li, C., Zhang, Z., Zhang, C., Zhang, Y., Qiu, J., Wang, Q., Liu, J., Jia, S. and Feng, J. (2015). Dynamical evolution of anisotropic response in black phosphorus under ultrafast photoexcitation, *Nano Lett.* 15, 4650.
46. Lange, S. and Nilges, T. (2006). Au3SnCuP10 and Au3SnP7: Preparation and crystal structures of Au3Sn heterocluster polyphosphides, *Z. Naturforsch. B.* 61, 871.
47. Nilges, T., Kersting, M. and Pfeifer, T. (2008). A fast low-pressure transport route to large black phosphorus single crystals, *J. Solid State Chem.* 181, 1707.
48. Wang, H., Yang, X., Shao, W., Chen, S., Xie, J., Zhang, X., Wang, J. and Xie, Y. (2015). Ultrathin black phosphorus nanosheets for efficient singlet oxygen generation, *J. Am. Chem. Soc.* 137, 11376.
49. Köpf, M., Eckstein, N., Pfister, D., Grotz, C., Krüger, I., Greiwe, M., Hansen, T., Kohlmann, H. and Nilges, T. (2014). Access and in situ growth of phosphorene-precursor black phosphorus, *J. Cryst. Growth.* 405, 6.
50. Nagao, M., Hayashi, A. and Tatsumisago, M. (2011). All-solid-state lithium secondary batteries with high capacity using black phosphorus negative electrode, *J. Power Sources* 196, 6902–6905.
51. Oumellal, Y., Rougier, A., Tarascon, J.M. and Aymard, L. (2009). 2LiH+M (M=Mg, Ti): New concept of negative electrode for rechargeable lithium-ion batteries, *J. Power Sources* 192, 698.
52. Park, C.M. and Sohn, H.J. (2007). Black phosphorus and its composite for lithium rechargeable batteries, *Adv. Mater.* 19, 2465–2468.
53. Sun, L.Q., Li, M.J., Sun, K., Yu, S.H., Wang, R.S. and Xie, H.M. (2012). Electrochemical activity of black phosphorus as an anode material for lithium-ion batteries, *J. Phys. Chem. C* 116, 14772–14779.
54. Li, Q.F., Duan, C.G., Wan, X.G. and Kuo, J.L. (2015). Theoretical prediction of anode materials in Li-ion batteries on layered black and blue phosphorus, *J. Phys. Chem. C* 119, 8662–8670.

55. Li, W., Yang, Y., Zhang, G. and Zhang, Y.M. (2015). Ultrafast and directional diffusion of lithium in phosphorene for high-performance lithium-ion battery, *Nano Lett.* 15, 1691–1697.

56. Zhao, S., Kang, W. and Xue, J. (2014). The potential application of phosphorene as an anode material in Li-ion batteries, *J. Mater. Chem. A* 2, 19046–19052.

57. Guo, G.C., Wang, D., Wei, X.L, Zhang, Q., Liu, H., Lau, W.M. and Liu, L.M. (2015). First-principles study of phosphorene and graphene hetero structure as anode materials for rechargeable Li batteries, *J. Phys. Chem. Lett.* 6, 5002–5008.

58. Sun, J., Zheng, G., Lee, H.W., Liu, N., Wang, H., Yao, H., Yang, W. and Cui, Y. (2014). Formation of stable phosphorus-carbon bond for enhanced performance in black phosphorus nanoparticle-graphite composite battery anodes, *Nano Lett.* 14, 4573–4580.

59. Wang, F. (2018). A dual-stimuli-responsive sodium-bromine battery with ultra-high energy density, *Adv. Mater.* 30, 1800028.

60. Wang, F., Wu, X., Li, C., Zhu, Y., Fu, L., Wu, Y. and Liu, X. (2016). Nanostructured positive electrode materials for post-lithium ion batteries, *Energy Environ. Sci.* 9, 3570–3611.

61. Kulish, V.V., Malyi, O.I., Perssoncd, C. and Wu, P. (2015). Phosphorene as an anode material for Na-ion batteries: A first-principles study, *Phys. Chem. Chem. Phys.* 17, 13921–13928.

62. Nie, A., Cheng, Y., Ning, S., Foroozan, T., Yasaei, P., Li, W., Song, B., Yuan, Y., Chen, L., Salehi-Khojin, A., Mashayek, F. and Yassar, R.S. (2016). Selective ionic transport pathways in phosphorene, *Nano Lett.* 16, 2240–2247.

63. Sun, J., Lee, H.W., Pasta, M., Yuan, H., Zheng, G., Sun, Y., Li, Y. and Cui, Y. (2015). A phosphorene-graphene hybrid material as a high-capacity anode for sodium-ion batteries, *Nat. Nanotechnol.* 10, 980–985.

64. Zhang, W., Mao, J., Li, S., Chen, Z. and Guo, Z. (2017). Phosphorus-based alloy materials for advanced potassium-ion battery anode, *J. Am. Chem. Soc.* 139, 3316–3319.

65. Zhu, Y.H., Yin, Y.B., Yang, X., Sun, T., Wang, S., Jiang, Y.S., Yan, J.M. and Zhang, X.B. (2017). Transformation of rusty stainless-steel meshes into stable, low-cost, and binder-free cathodes for high-performance potassium-ion batteries, *Angew. Chem. Int. Ed.* 56, 7881–7885.

66. Su, D., McDonagh, A., Qiao, S.Z. and Wang, G. (2017). High-capacity aqueous potassium-ion batteries for large-scale energy storage, *Adv. Mater.* 29, 1604007.

67. Chen, Z., Yin, D. and Zhang, M. (2018). Sandwich-like MoS_2@ SnO_2@ C with high capacity and stability for sodium/potassium ion batteries, *Small* 14, 1703818.

68. Lei, K., Wang, C., Liu, L., Luo, Y., Mu, C., Li, F. and Chen, J. (2018). A porous network of bismuth used as the anode material for high-energy-density potassium-ion batteries, *Angew. Chem. Int. Ed.* 130, 4687–4691.

69. Xu, Y., Zhang, C., Zhou, M., Fu, Q., Zhao, C., Wu, M. and Lei, Y. (2018). Highly nitrogen doped carbon Nano fibers with superior rate capability and cyclability for potassium ion batteries, *Nat. Commun.* 9, 1720.

70. Sultana, I., Rahman, M.M., Ramireddy, T., Chen, Y. and Glushenkov, A.M. (2017). High capacity potassium-ion battery anodes based on black phosphorus, *J. Mater. Chem. A* 5, 23506–23512.

71. Wu, X., Zhao, W., Wang, H., Qi, X., Xing, Z., Zhuang, Q. and Ju, Z. (2018). Enhanced capacity of chemically bonded phosphorus/carbon composite as an anode material for potassium-ion batteries, *J. Power Sources* 378, 460–467.

72. Pang, Q., Liang, X., Kwok, C.Y. and Nazar, L.F. (2016). Advances in lithium-sulphur batteries based on multifunctional cathodes and electrolytes, *Nat. Energy* 1, 116132.

73. Xu, J., Lawson, T., Fan, H. Su, D. and Wang, G. (2018). Updated metal compounds (MOFs, S, OH, N, C) used as cathode materials for lithium-sulphur batteries, *Adv. Energy Mater.* 8, 1702607.

74. Chen, W., Lei, T., Wu, C., Deng, M., Gong, C., Hu, K., Ma, Y., Dai, L., Lv, W., He, W., Liu, X., Xiong, J. and Yan, C. (2018). A new hydrophilic binder enabling strongly anchoring poly sulphides for high-performance sulphur electrodes in lithium-sulphur battery, *Adv. Energy Mater.* 8, 1702348.

75. Dong, R., Pfeffermann, M., Skidin, D., Wang, F., Fu, Y., Narita, A., Tommasini, M., Moresco, F., Cuniberti, G., Berger, R., Müllen, K. and Feng, X. (2017). Persulfurated coronene: A new generation of "sulflower", *J. Am. Chem. Soc.* 139, 2168–2171.

76. Li, L., Chen, L., Mukherjee, S., Gao, J., Sun, H., Liu, Z., Ma, X., Gupta, T., Singh, C.V., Ren, W. and Cheng, H.M. (2017). Phosphorene as a polysulfide immobilizer and catalyst in high-performance lithium-sulphur batteries, *Adv. Mater.* 29, 1602734.

77. Jin, W., Wang, Z. and Fu, Y.Q. (2016). Monolayer black phosphorus as potential anode materials for Mg-ion batteries, *J. Mater. Sci.* 51, 7355–7360.

12 Application and Performance Analysis of Various Nature-Inspired Algorithms in AGC Synthesis

Shelja
Tilak Raj Chadha Institute of Management
and Technology (TIMT)
NIILM University

Pawan Kumar
NIILM University

Aman Ganesh
Lovely Professional University

CONTENTS

DOI: 10.1201/9781003242277-12

12.1 INTRODUCTION

Electric power systems are interconnected systems that are always required to meet the ever-growing consumer electric power demand without compromising both the quality and reliability of the supplied power. This operation becomes even more complex under the constraints of (i) less dependency on fossil fuel-based power owing to the exhaustible nature of the fuel and growing environmental concerns, (ii) growing penetration of renewable energy resources for economic operation because of the inexpensive and inexhaustible nature of the fuels such as wind and solar, and (iii) more stressed transmission lines, which are made to operate near thermal limits due to socio-economic constraints on expanding transmission network.

For the successful operation of the interconnected power systems, it is AGC that continually balances the load–generation pattern in such a way that it results in minimizing the area control error which has direct bearing on minimization of frequency deviation. This is complemented by changing the tie line power so that a balance between load and generation is always maintained. All this is generally examined using real-time simulations [1].

AGC synthesis, operation, and control have a voluminous and tremendous history, starting from the dynamic control of the system nonlinearities, viz. governor dead band, generation ramp-up time, mapping of load characteristics followed by the interaction between the active and reactive power seen in terms of frequency and voltage control using the AGC, and the AVR loop control.

But in last few decades, the research has shifted focus in dealing with the increased penetration of the renewable energy sources such as inclusion of wind- and solar-based power generation units and energy storage devices.

For a real-time system with the availability of the deterministic model, many linear and nonlinear approaches such as optimal control, robust control, and adaptive control have been presented and implemented. But for complex power systems with insufficient knowledge, intelligent control techniques based on many heuristic and metaheuristic algorithms have been presented [2]. The control problem mostly involved optimizing the cost functions so as to minimize the frequency deviation and area control error, which are two major objectives of the AGC control problem. The optimized problem largely relates to the parametric optimization of the AGC regulator, which is generally a PID controller. The optimization problem helps in finding the global solution so that the AGC regulator is able to enhance the AGC insensitivity to the parametric or the system variations.

Although many optimization techniques [3–25] have been employed, controller tuning method has not been explicitly explained. No general procedure has been advocated about the use or implementation of the optimization algorithm. The present work investigates the state-of-the-art optimization of different performance metrics in a multi-area network so that a general procedure may be adopted for the selection of the optimization algorithm based on the parameters such as the number of search agents involved, speed, and accuracy of the algorithm subject to time constraints. Since the AGC forms the integral part of the ancillary services and emerging electricity market, a fast and best result helps in reducing the frequency deviation under the effect of major disturbances.

12.2 MODERN AGC

This section introduces the structure of AGC where the mechanism revolves around minimizing the frequency deviation and regulating tie line power flow in the interconnected power system. The frequency deviation is due to the imbalance between the generation and load, and the frequency of the electrical network is proportional to the generator's rotating speed. Hence, the frequency control problem is directly translated to speed control of the generator where the speed-governing mechanism is involved which senses the machine speed and regulates the mechanical output power for tracking the load change and hence minimizing the frequency deviation by restoring the frequency to the nominal value. Figure 12.1 shows the same mechanism followed in each area for the shown two-area power network.

A very significant issue related to AGC control is the response of the generating unit to the type of the fuel, the unit type, and the operating point. The unit type plays a vital role since the dynamics of the fossil fuel-based thermal unit are different from the hydropower unit, which is altogether different from the renewable energy-based generating units such as wind and solar energy storages. All have different inertia and different ramp-up rates. Hence, the AGC performance is highly system dependent.

The other aspect is the control strategy related to AGC. Conventionally, integral controllers were used in the AGC loop so as to bring the frequency back to the nominal value following the load–generation imbalance. Now, since modern power systems are multi-generation multi-area networks, controller types and the adopted tuning methods have a significant impact on the AGC performance. Usually, the

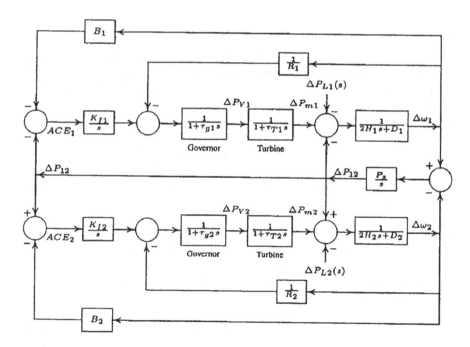

FIGURE 12.1 Basic AGC structure.

proportional–integral–derivative (PID) controller in any of the combinations as integral (I), proportional–integral (PI), or PID controller is used. Since PID controllers are linear controllers, it becomes very important to have the prior knowledge of the power system so that controller parameter tuning can be done. But due to restructuring and deregulation many a times, the complete information about the power system is not available; hence, controller tuning draws maximum attention. Many heuristic and metaheuristic algorithms have been used for determining the optimal value of the controller parameters.

12.3 OPTIMIZATION METHODS

Optimization algorithms are basically search algorithms that are intended to provide the best feasible/possible solution from the available set of solutions. The heuristic algorithms have the problem awareness, and they depend upon the quality of the solution provided in each iteration, which always connects the proximity of the past, current, and target solutions. Greedy search and hill climbing are well-known examples of heuristic algorithms. Metaheuristic algorithms treat the problem as a black box and explore the whole search space by using the stochastic approach, thus finding an optimal solution. Although the list is exhaustive, genetic algorithm and simulated annealing are the popular examples of metaheuristic algorithms.

In this chapter, advanced optimization algorithms such as grey wolf optimization, particle swarm optimization, salp swarm algorithm, and whale optimization algorithm are explored.

12.3.1 GREY WOLF OPTIMIZATION

Grey wolf optimizer (GWO) is a population-based metaheuristics algorithm that simulates the strict social dominant hierarchy and hunting mechanism of gray wolves. It was proposed by Mirjalili et al. in 2014. Grey wolves are at the top of the food chain because they are considered apex predators. They prefer to live in a group of size varying from 5 to 12. The hierarchy has four levels depicting the four types of grey wolves, namely alpha, beta, delta, and omega. The alpha (α) level depicts the leadership, where the alpha male or female is responsible for making decisions. The beta (β) level depicts the subordination wolves, which help in decision making and enforcing alpha's command along with providing feedback to the alpha wolf. The delta (δ) wolves are third in the hierarchy and are responsible for watching the boundary and raising alarm in case of any danger. The omega (ω) level wolves are the scapegoats, but the whole pack fights in case of losing the omega. The major steps of hunting are followed, which are searching which involves tracking, chasing, and approaching the prey; encircling which usually also includes pursuing and harassing the prey until its stops moving; and lastly attacking the prey. The hunting is guided by α, β, and δ, while ω follows these three wolves.

The encircling behavior is mathematically modeled as

$$D = |C \cdot X_p(t) - X(t)| \tag{12.1}$$

$$X(t+1) = X_p(t) - A \cdot D \tag{12.2}$$

where X_p indicates the position vector of the prey, X is the position vector of a grey wolf, t is the current iteration, and A and C are coefficient vectors and are calculated as follows:

$$A = 2a \cdot r_1 - a \tag{12.3}$$

$$C = 2 \cdot r_2 \tag{12.4}$$

where the components of a are linearly decreased from 2 to 0 over the course of iterations and r_1, r_2 are random vectors in $[0, 1]$.

12.3.2 PARTICLE SWARM OPTIMIZATION

The particle swarm optimization (PSO) is a nature-inspired metaheuristic algorithm that depicts the social behavior and dynamic movements with communications of particles within a swarm. The examples include ant colony, bee colony, and bird flocking. Here, ants, bees, and birds are the particles, whereas colonies and flock represent the swarm. PSO works on three basic principles of separation, alignment, and cohesion. These aspects combine the self and the social experience of the individuals. Here, the collection of particles forms the changing solution; search area is the possible solution. The movement of the particles to toward a promising area is to get the global optimum solution. In a swarm, each particle keeps the track of the personal best (p_{best}) and global best solution (g_{best}). The solution lies in the fact that each particle in the group modifies its position according to its current position and velocity and distance between its personal best and global best positions.

The algorithm works by initializing all the particles of the swarm with random position and velocity. Then the particle's position and velocity by checking the fitness for the defined p_{best} and g_{best} are updated. Mathematically, it is represented as follows:

$$v_i = Wv_i + c_1 r_1 \left(P_{\text{best},i} - x_i \right) + c_2 r_2 \left(g_{\text{best}} - x_i \right) \tag{12.5}$$

$$x_i = x_i + v_i \tag{12.6}$$

where i is the ith particle, v and x are the respective velocity and position of the particle, r_1 and r_2 are the random numbers, c_1 and c_2 are the acceleration coefficients, and W is the inertia weight.

12.3.3 SALP SWARM OPTIMIZATION ALGORITHM

The salp swam algorithm (SSA) is a type of multiobjective nature-inspired optimizer. In deep oceans, salps form a swarm chain for achieving better locomotion and foraging with rapid change in coordinates. The swarm is divided into leader and follower groups for the search of the food. The salp leader guides the swarm or the follower. Based on the food source (F), the leader updates its position. The position of the leader salp x in the jth dimension is evaluated in terms of the food position F. The solution set is range bound with lb the lower bound and ub the upper bound, each of jth dimension. Mathematically, SSA is represented as follows:

$$x_j^1 = \begin{cases} F_j + c_1\big((\text{ub}_j - \text{lb}_j)c_2 + \text{lb}_j\big) & c_3 \geq 0 \\ F_j - c_1\big((\text{ub}_j - \text{lb}_j)c_2 + \text{lb}_j\big) & c_3 < 0 \end{cases} \tag{12.7}$$

Here, c_1 the most important coefficient that helps in exploration and exploitation and is defined as follows:

$$c_1 = 2e^{-\left(\frac{4I}{L}\right)^2} \tag{12.8}$$

where I and L represent the current and the maximum number of iterations, respectively.

c_2 and c_3 are random numbers, which are uniformly generated in the interval of $[0,1]$.

The follower position is updated using the following equation:

$$x_j^i = \frac{1}{2}at^2 + v_o t \tag{12.9}$$

Equation (12.9) holds good for all is greater than or equal to 2, and here, the position of the ith follower salp is shown in jth dimension with t as time (iteration in optimization) and v_o as the initial velocity.

12.3.4 WHALE OPTIMIZATION ALGORITHM

Whale optimization is also a nature-inspired swarm intelligence-based algorithm that mimics the hunting behavior of humpback whales. The presence of the spindle cells makes whale to think, learn, judge, and communicate. The humpback whale exhibits bubble-net feeding method for foraging. These whales create a spiral-shaped bubble net around the prey in three stages, namely coral loop, lobtail, and capture loop. The coral loop or the encircling prey mechanism is mathematically represented as follows.

The encircling behavior is mathematically modeled as shown below:

$$D = |C \cdot X_p(t) - X(t)| \tag{12.10}$$

$$X(t+1) = X_p(t) - A \cdot D \tag{12.11}$$

where X_p indicates the position vector of the prey, X is the position vector of a grey wolf, t is the current iteration, and A and C are coefficient vectors and are calculated as

$$A = 2a \cdot r_1 - a \tag{12.12}$$

$$C = 2 \cdot r_2 \tag{12.13}$$

Equations (12.11) and (12.12) are together used to give a 3D view of position (X, Y) with depth r.

The algorithm works first by shrinking the encirclement by decreasing the value of a and hence the space vector A followed by updating the position so a spiral is created between the whale and the prey the mathematical model is represented as helix and as depicted as

$$\vec{X}(t+1) = \vec{D}' \cdot e^{bl} \cdot \cos(2\pi l) + \vec{X}^*(t) \tag{12.14}$$

where $\vec{D}' = \left| \vec{X}^*(t) - \vec{X}(t) \right|$ indicates the distance between the ith whale and the prey and in optimization is termed as the best solution (so far), b is a constant and defines the spiral, and l represents a random number ranging between -1 and 1. So with p as a probability factor ranging between 0 and 1 with 50% chance as a breakaway solution, the complete mathematical model is given as

$$\vec{X}(t+1) = \begin{cases} \vec{X}^*(t) - \vec{A} \cdot \vec{D} & \text{if } p < 0.5 \\ \vec{D}' \cdot e^{bl} \cdot \cos(2\pi l) + \vec{X}^*(t) & \text{if } p > 0.5 \end{cases} \tag{12.15}$$

12.4 MODEL DESCRIPTION

The considered multi-area network is also a multi-input network. A two-area hybrid model is considered, which consists of conventional generation involving steam-based generation (SG) along with renewable generation involving solar photovoltaic system (SPV) and wind energy-based generation system (WG). Each area consists of coupled units, each having the rating of 900 MVA, 24 kV at 50Hz. The two areas have the speed regulation of 0.05 and 0.0625 with the frequency-sensitive load deviation of 0.6 and 0.9, respectively. The governor and turbine time constants for the two areas are [0.2 s, 0.5 s] and [0.3 s, 0.6 s], respectively. The solar time constant was 1.8 s, and the solar gain constant was 1.0 for both the areas. The wind constant was 1, and the wind time constant was 0.8 for both the areas. The common base power is 1,000 MVA. 10% load variation is considered. The AGC performance is studied using the PID controller, and the four metaheuristic algorithms listed are used to find the optimized controller parameters.

12.4.1 CONTROLLER IMPLEMENTATION

Metaheuristic algorithms are employed to find the optimal gain parameters, i.e., the proportional gain (K_p) and integral gain (K_i). The combined controller output is defined as

$$G_c(s) = K_p + \frac{K_i}{s} \tag{12.16}$$

One of the widely employed objective functions for obtaining the control parameters of AGC is the time integral of absolute error (ITAE) based on frequency deviation of both the interconnected areas defined as Δf_1, Δf_2, and deviation in tie line power ΔP_{12}. The formulated objective function is expressed as

$$J = \int \left(|\Delta f_1| + |\Delta f_2| + |\Delta P_{21}| \right) dt \tag{12.17}$$

The optimization algorithms are tested to minimize the objective function under the controller parameter constraints (K_p and K_i) defined as

$$K_{pj}(\min) \leq K_{pj} \leq K_{pj}(\max) \text{ and } K_{ij}(\min) \leq K_{ij} \leq K_{ij}(\max) \tag{12.18}$$

where j has the value 1 or 2 corresponding to area 1 and 2, respectively.

12.5 RESULTS

The load frequency control is the major function of automatic generation control. The controller operates so as to regulate the power generation, resulting in minimizing the frequency deviation. The PID-based AGC regulator controls the participation index of the involved generating units based on the regulator control signal, which in turn is the function of the controller parameters. For the sudden imbalance created between generation and load due to the sudden change in load demand, the system frequency experiences a transient change. During this transient period, there is power flow through the tie line to the area experiencing higher power demand compared to the other areas. So under the steady condition, both the tie line power and frequency deviation will fall to zero value, indicating the balance match between load and generation.

This section gives the comparative performance analysis of the used optimization techniques, namely GWO, PSO, SSA, and WOA. The obtained controller parametric values are given in Table 12.1.

Based on the AGC regulator parametric value of the PI controller, the frequency deviation of area 1 is shown in Figure 12.2.

From the obtained results, it is evident that the GWO shows better efficacy in achieving load–generation balance, which is exhibited by its ability to obtain zero frequency deviation in least time in comparison with the other three selected optimizers.

TABLE 12.1
Obtained Tuned Parameters Using
Different Algorithms

Algorithm	Ki1	Ki2	Kp1	Kp2
GWO	1.818	1.938	0.605	0.127
SSA	3.500	2.015	1.249	1.529
PSO	0.773	0.148	0.151	0.286
WOA	0.882	0.505	1.235	4.506

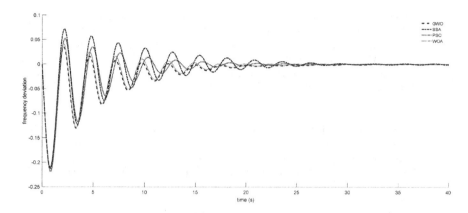

FIGURE 12.2   Frequency deviation of area 1.

12.6   CONCLUSIONS

This chapter presented a systematic approach to comparing the performance of select nature-inspired optimization algorithms for optimally synthesizing the automatic generation control employed in the multi-area power network. The grey wolf optimizer exhibited the best performance in a defined multi-area multisource network. The comparison is made between grey wolf optimization, particle swarm optimization, salp swarm algorithm, and whale optimization algorithm.

REFERENCES

1. H. Saadat (2014). *Power System Analysis*, McGraw-Hill Series in Electrical and Computer Engineering, New York.
2. S. Mirjalili and J.S. Dong (2019). *Multi Objective Optimization Using Artificial Intelligence Techniques*, Springer Nature, Switzerland.
3. O.I. Elgerd and C.E. Fosha (1970). Optimum megawatt-frequency control of multi area electric energy systems. *IEEE Transactions on Power Apparatus and Systems*, 89(4), 556–563,
4. N. K. Kumar and V. Indragandhi (2018). Analysis on various optimization techniques used for load frequency control in power system. *Serbian Journal of Electrical Engineering*, 15(3), 249–273.
5. P. Ojaghi and M. Rahmani (2017). LMI-based robust predictive load frequency control for power systems with communication delays. *IEEE Transactions on Power Systems*, 32(5), 4091–4100.
6. V. Gholamrezaie, M. G. Dozein, H. Monsef and B. Wu (2018). An optimal frequency control method through a dynamic Load Frequency Control (LFC) model incorporating wind farm. *IEEE Systems Journal*, 12(1), 392–401.
7. Y. Xu, C. Li, Z. Wang, N. Zhang and B. Peng (2018). Load frequency control of a novel renewable energy integrated micro-grid containing pumped hydropower energy storage. *IEEE Access*, 6, 29067–29077.
8. Y. Zhang and T. Yang (2020). Decentralized switching control strategy for load frequency control in multi-area power systems with time delay and packet losses. *IEEE Access*, 8, 15838–15850.

9. H. Bevrani and P. R. Daneshmand (2012). Fuzzy logic-based load-frequency control concerning high penetration of wind turbines. *IEEE Systems Journal*, 6(1), 73–180.

10. H. A. Yousef, K. AL-Kharusi, M. H. Albadi and N. Hosseinzadeh (2014). Load frequency control of a multi-area power system: An adaptive fuzzy logic approach. *IEEE Transactions on Power Systems*, 29(4), 1822–1830.

11. T. Fernando, K. Emami, S. Yu, H. H. Iu and K. P. Wong (2016). A novel quasi-decentralized functional observer approach to LFC of interconnected power systems. *IEEE Transactions on Power Systems*, 31(4), 3139–3151.

12. N. E. Y. Kouba, M. Menaa, M. Hasni and M. Boudour (2016). LFC enhancement concerning large wind power integration using new optimised PID controller and RFBs. *IET Generation, Transmission & Distribution*, 10(16), 4065–4077.

13. I. Nasiruddin, G. Sharma, K. R. Niazi and R. C. Bansal (2017). Non-linear recurrent ANN-based LFC design considering the new structures of Q matrix. *IET Generation, Transmission & Distribution*, 11(11), 2862–2870.

14. X. Wang, Q. Zhao, B. He, Y. Wang, J. Yang and X. Pan (2017). Load frequency control in multiple microgrids based on model predictive control with communication delay. *The Journal of Engineering*, 17(13), 1851–1856.

15. D. Qian and G. Fan (2018). Neural-network-based terminal sliding mode control for frequency stabilization of renewable power systems. *IEEE/CAA Journal of Automatica*, 5(3), 706–717.

16. I. Nasiruddin, G. Sharma, K.R. Niazi and R.C. Bansal (2018). Non-linear recurrent ANN-based LFC design considering the new structures of Q matrix. *IET Generation, Transmission & Distribution*, 12(16), 3912–3912.

17. Z. Yan and Y. Xu (2019). Data-driven load frequency control for stochastic power systems: A deep reinforcement learning method with continuous action search. *IEEE Transactions on Power Systems*, 34(2), 1653–1656.

18. C. Osinski, G. Villar Leandro and G. H. da Costa Oliveira (2019). Fuzzy PID controller design for LFC in electric power systems. *IEEE Latin America Transactions*, 17(01), 147–154.

19. J. Yang, X. Sun, K. Liao, Z. He and L. Cai (2019). Model predictive control-based load frequency control for power systems with wind-turbine generators. *IET Renewable Power Generation*, 13(15), 2871–2879.

20. H. Haes Alhelou, M. E. H. Golshan and N. D. Hatziargyriou (2019). A decentralized functional observer based optimal LFC considering unknown inputs, uncertainties, and cyber-attacks. *IEEE Transactions on Power Systems*, 34(6), 4408–4417.

21. C. J. Ramlal, A. Singh, S. Rocke and M. Sutherland (2019) Decentralized fuzzy H∞ iterative learning LFC with time-varying communication delays and parametric uncertainties. *IEEE Transactions on Power Systems*, 34(6) 4718–4727.

22. Z. Cheng, D. Yue, S. Hu, X. Xie and C. Huang (2019). Detection-based weighted H∞ LFC for multi-area power systems under DoS attacks. *IET Control Theory & Applications*, 13(12), 1909–1919.

23. S. Aziz, H. Wang, Y. Liu, J. Peng and H. Jiang (2019). Variable universe fuzzy logic-based hybrid LFC control with real-time implementation. *IEEE Access*, 7, 25535–25546.

24. D. Zhou, Z. Quan and Y. Li (2020). Hybrid model predictive control of ANPC converters with decoupled low-frequency and high-frequency cells. *IEEE Transactions on Power Electronics*, 35(8), 8569–8580.

25. D. Yang, et al. (2020). Inertia-adaptive model predictive control-based load frequency control for interconnected power systems with wind power. *IET Generation, Transmission & Distribution*, 14(22), 5029–5036.

13 Unified Smith Predictor for MIMO Systems with Multiple Time Delays

Ashu Ahuja
M. M. University

Bhawna Tandon
Chandigarh Engineering College

CONTENTS

13.1   INTRODUCTION

In this work, a design methodology is developed to design a 2-DOF (degree of freedom) controller for multi-input multi-output (MIMO) systems having multiple time delays. MIMO systems with multiple delays are converted to lower-fractional transformation using weighting function whose parameters are tuned using a genetic algorithm (GA). The converted delayed plant is transformed to augmented plant using the unified Smith predictor (USP). Setpoint tracking is isolated from disturbance rejection using the 2-DOF controller structure. An observer-based feedback controller is designed for disturbance rejection and reference-based model matching controller for setpoint tracking. Robust stability is obtained from the designed controller.

13.2   LITERATURE SURVEY

Time delays appear in many physical systems, especially in those involving information transmissions. The most famous method to deal with such kind of systems is the

DOI: 10.1201/9781003242277-13

Smith predictor (SP). The SP designs the controller in such a way that the closed-loop response of a delayed system is the same as the delayed response of a closed-loop delay-free system [1,2]. The SP is also applied to time-varying delay [3]. But, it can only be applied to stable systems in which delay is perfectly known. This is the limitation over SP as many delayed systems are unstable.

Watanabe and Ito [4] presented a method called process model control for general unstable process with dead time. Matausek and Micic presented a modified Smith predictor (MSP) by firstly introducing a minor loop to stabilize the process with a proportional gain [5] and later another scheme with a high-pass filter [6]. Majhi and Atherton [7] presented a scheme for stable/unstable processes with dead time. However, a delay element still exists in the characteristics equation of the disturbance response in the scheme studied in [5–7]. Mirkin and Zhong [8] and Alcantra [9] proposed a scheme that acts as a general framework to unify dead time compensation-based control of stable and unstable processes. Febina Beevi applied approximate generalized time moments (AGTM) matching concept to design a 2-DOF controller for integral [10] and first-order [11] time delay systems. Other 2-DOF controllers are also designed for time-delayed systems such as predictor-based controller [12,13], observer-based controller [14], model reference controller [15], input-/output-linearized and high-gain controller [16], fictitious reference iterative tuning controller [17], and Pareto-based polynomial tuning-ruled PID controller [18]. The work is extended to MIMO systems with delay. MIMO systems may have a single delay [19,20] or multiple delays in multiple channels [21–23].

But, all of these techniques are applied to either stable or unstable systems. If the systems have both kinds of poles—stable and unstable, then SP and MSP can't be applied. SP gives poor robustness, and it is difficult to ensure a minimum damping ratio of the closed-loop system when the open-loop system has poorly damped poles, and in case of systems having fast, stable eigenvalues, the MSP algorithms may be numerically unstable. Then the USP was proposed [24], which does not require matrix exponential computation for fast, stable poles, and it was applied to power system damping control [25].

But, because of interaction among various loops, it becomes difficult to design controller for MIMO systems with multiple delays. Decoupling is an effective tool to eliminate the aforementioned obstacles. This method decouples the MIMO process into several independent SISO systems [26,27]. A robust, fixed full-order matrix H_∞ controller is designed for MIMO systems with multiple delays by convex approximation in the frequency domain by a set of complex values in [28–30].

In the present work, the USP design methodology is extended to MIMO systems with multiple time delays. This chapter is organized as follows:

Section 13.2 describes the design methodology of the augmented plant using the USP. Section 13.3 explains the parameterization of the 2-DOF controller structure. Section 13.4 implements the designed approach to a two-input two-output MIMO system, and Section 13.5 concludes this chapter.

13.3 UNIFIED SMITH PREDICTOR FOR MULTIPLE TIME DELAYS

A system having multiple time delays is represented in Figure 13.1.

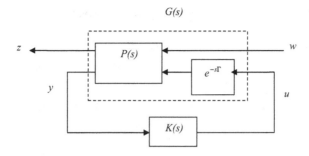

FIGURE 13.1 Control system comprising multiple time-delayed plants $G(s)$ and controllers $K(s)$.

The system $G(s)$ has "n" number of delayed inputs. So, "u" is a vector of components $[u_1, u_2, \ldots, u_n]$, which are delayed by τ_1, τ_2, τ_3, $\ldots$, $\tau_n \geq 0$. The delayed components are represented by a matrix $\Gamma = \mathrm{diag}(\tau_1, \tau_2, \ldots, \tau_n)$. So, the system with multiple delays is represented by

$$G(s) = P(s)e^{-s\Gamma} \tag{13.1}$$

where $P(s)$ is a delay-free augmented plant including weights and its state space representation is given as

$$
P(s) = \left[\begin{array}{ccc}
A & \vdots & B_1 \quad B_2 \\
\cdots & \vdots & \cdots \quad \cdots \\
C_1 & \vdots & D_{11} \quad D_{12} \\
C_2 & \vdots & D_{21} \quad D_{22}
\end{array} \right] = \left[\begin{array}{cc}
P_{11} & P_{12} \\
P_{21} & P_{22}
\end{array} \right]
$$

The delay-free part $P(s)$ is decomposed into sum of stable part P_s and unstable part P_u. The classical Smith predictor (CSP) is designed for P_s, and MSP for P_u. The linear coordinate transformation in its state space representation is used to decompose $P(s)$. The transformation matrix, V, is obtained using MATLAB® function "eig", and the elements of transformed matrix are converted from complex diagonal form to real diagonal form using MATLAB function "cdf2rdf". $[V, D] = \mathrm{cdf2rdf}(V, D)$. A_u and A_s are the unstable and stable parts of A after transforming into Jordan canonical form. This decomposition is made by splitting the complex plane along with a vertical line $\mathrm{Re}(s) = \sigma$ with $\sigma < 0$. The value of σ is chosen as the maximum negative real part of poorly damped poles. Then the eigenvalues of A_u are all eigenvalues $\{\lambda_i\}$ of A with $\mathrm{Re}(\lambda_i) > \sigma$, while A_s has remaining eigenvalues of A.

The transformed augmented delay-free plant P_{22}^t is obtained as follows:

$$
P_{22}^t(s) = \left[\begin{array}{ccc}
V^{-1}AV & \vdots & V^{-1}B_2 \\
\cdots & \vdots & \cdots \\
C_2V & \vdots & 0
\end{array} \right] = \left[\begin{array}{cccc}
A_u & 0 & \vdots & B_u \\
0 & A_s & \vdots & B_s \\
\cdots & \cdots & \vdots & \cdots \\
C_u & C_s & \vdots & 0
\end{array} \right] \tag{13.2}
$$

Here, P_{22}^t is the sum of stable and unstable parts of the plant.

$$P_{22}^t = P_u + P_s \tag{13.3}$$

where

$$P_u = \begin{bmatrix} A_u & \vdots & B_u \\ \cdots & \vdots & \cdots \\ C_u & \vdots & 0 \end{bmatrix} \text{ and } P_s = \begin{bmatrix} A_s & \vdots & B_s \\ \cdots & \vdots & \cdots \\ C_s & \vdots & 0 \end{bmatrix} \tag{13.4}$$

The classical Smith predictor is considered for P_s and represented by $Z_s(s)$ as follows:

$$Z_s(s) = P_s(s) - P_s(s)e^{-s\Gamma} \tag{13.5}$$

and the modified Smith predictor is considered for P_u and represented by $Z_u(s)$:

$$Z_u(s) = P_u^{\text{aug}}(s) - P_u(s)e^{-s\Gamma} \tag{13.6}$$

where

$$P_u^{\text{aug}}(s) = \begin{bmatrix} A_u & \vdots & B_u \\ \cdots & \vdots & \cdots \\ C_u e^{-A_u \Gamma} & \vdots & 0 \end{bmatrix} = \begin{bmatrix} A_u & \vdots & e^{-A_u \Gamma} B_u \\ \cdots & \vdots & \cdots \\ C_u & \vdots & 0 \end{bmatrix} \tag{13.7}$$

Let $B_u = \begin{bmatrix} \beta_1 & \beta_2 & \dots & \beta_n \end{bmatrix}$ and $C_u = \begin{bmatrix} \varsigma_1 \\ \varsigma_2 \\ \vdots \\ \varsigma_n \end{bmatrix}$, then

$$e^{-A_u \Gamma} B_u = \begin{bmatrix} e^{-A_u \tau_1} \beta_1 & e^{-A_u \tau_2} \beta_2 & \dots & e^{-A_u \tau_n} \beta_n \end{bmatrix}$$

$$C_u e^{-A_u \Gamma} = \begin{bmatrix} \varsigma_1 e^{-A_u \tau_1} \\ \varsigma_2 e^{-A_u \tau_2} \\ \vdots \\ \varsigma_n e^{-A_u \tau_n} \end{bmatrix} \tag{13.8}$$

The USP, $Z(s)$, is the sum of CSP ($Z_s(s)$) and MSP ($Z_u(s)$) and is given as

$$Z(s) = Z_s(s) + Z_u(s)$$

$$= P_{22}^{\text{aug}}(s) - P_{22}^t(s)e^{-s\Gamma} \tag{13.9}$$

where $P_{22}^{\text{aug}} = P_s + P_u^{\text{aug}}$ and it is realized as

$$P_{22}^{\text{aug}}(s) = \begin{bmatrix} A & \vdots & B_2 \\ \cdots\cdots & \vdots & \cdots\cdots \\ C_2 E_\Gamma & \vdots & 0 \end{bmatrix} \tag{13.10}$$

where

$$E_\Gamma = V \begin{bmatrix} e^{-A_u \Gamma} & 0 \\ 0 & I_s \end{bmatrix} V^{-1} \tag{13.11}$$

and

$$C_2 E_\Gamma = \begin{bmatrix} C_2(1,:)E_{\tau_1} \\ C_2(2,:)E_{\tau_2} \\ \vdots \\ C_2(n,:)E_{\tau_n} \end{bmatrix} \tag{13.12}$$

$E_{\tau_1}, E_{\tau_2}, \ldots, E_{\tau_n}$ can be expressed as

$$E_{\tau_1} = V \begin{bmatrix} e^{-A_s \tau_1} & 0 \\ 0 & I_s \end{bmatrix} V^{-1}; \quad E_{\tau_2} = V \begin{bmatrix} e^{-A_s \tau_2} & 0 \\ 0 & I_s \end{bmatrix} V^{-1}; \quad E_{\tau_n}$$

$$= V \begin{bmatrix} e^{-A_s \tau_n} & 0 \\ 0 & I_s \end{bmatrix} V^{-1} \tag{13.13}$$

Here, I_s is the identity matrix having the same dimensions as A_s. The new augmented plant ($\tilde{P}(s)$) is obtained by connecting the original plant and the USP in parallel as shown in Figure 13.2.

$$\tilde{P}(s) = \begin{bmatrix} P_{11}(s) & P_{12}(s)e^{-s\Gamma} \\ P_{21}(s) & P_{22}^{\text{aug}}(s) \end{bmatrix} \tag{13.14}$$

where

$$P_{11}(s) = \begin{bmatrix} \begin{bmatrix} A_u & 0 \\ 0 & A_s \end{bmatrix} & \vdots & \begin{bmatrix} 0 & 0 \\ 0 & e^{A_s \Gamma} - I_s \end{bmatrix} V^{-1} B_1 \\ \cdots\cdots\cdots & \vdots & \cdots\cdots\cdots\cdots\cdots \\ C_1 V & \vdots & 0 \end{bmatrix} \tag{13.15}$$

$$P_{12}(s) = \begin{bmatrix} A & \vdots & B_2 \\ \cdots & \vdots & \cdots \\ C_1 & \vdots & D_{12} \end{bmatrix} \qquad (13.16)$$

$$P_{21}(s) = \begin{bmatrix} A & \vdots & E_\Gamma^{-1} B_1 \\ \cdots\cdots & \vdots & \cdots\cdots\cdots \\ C_2 E_\Gamma & \vdots & D_{21} \end{bmatrix} \qquad (13.17)$$

$$P_{22}^{aug} = \begin{bmatrix} A & \vdots & B_2 \\ \cdots\cdots & \vdots & \cdots \\ C_2 E_\Gamma & \vdots & 0 \end{bmatrix} \qquad (13.18)$$

The generalized plant obtained by combining (13.15) with (13.18) is shown in Figure 13.2. $\tilde{P}(s)$ can be written as

$$\tilde{P} = \begin{bmatrix} A & 0 & \vdots & E_\Gamma^{-1} B_1 & B_2 \\ 0 & A_s & \vdots & \begin{bmatrix} 0 & e^{A_s \Gamma} - I_s \end{bmatrix} V^{-1} B_1 & 0 \\ \cdots\cdots & \cdots\cdots & \vdots & \cdots\cdots\cdots\cdots & \cdots\cdots \\ C_1 & C_1 V \begin{bmatrix} 0 \\ I_s \end{bmatrix} & \vdots & 0 & D_{12} \\ C_2 E_\Gamma & 0 & \vdots & D_{21} & 0 \end{bmatrix}$$

$$= \begin{bmatrix} P_a & \vdots & P_{b_1} & P_{b_2} \\ \cdots\cdots & \vdots & \cdots\cdots\cdots \\ P_{c_1} & \vdots & 0 & P_{d_{12}} \\ P_{c_2} & \vdots & P_{d_{21}} & 0 \end{bmatrix}$$

$$(13.19)$$

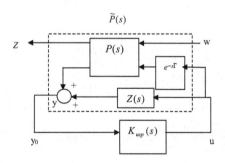

FIGURE 13.2   Plant $G(s)$ together with USP and controller $K_{usp}(s)$.

The expressions used in $\tilde{P}$ are defined as

$$E_\Gamma^{-1} B_1 = \begin{bmatrix} E_{\tau_1}^{-1} B_1(:,1) & E_{\tau_2}^{-1} B_1(:,2) & \dots & E_{\tau_n}^{-1} B_1(:,n) \end{bmatrix} \quad (13.20)$$

and

$$[0e^{A_s\Gamma} - I_s]V^{-1}B_1 = [[0e^{A_s\tau} - I_s]V^{-1}B_1(:,1)$$
$$[0e^{A_sA\tau_2} - I_s]V^{-1}B_1(:,2)\dots[0e^{A_s\tau n} - I_s]V^{-1}B_1(:,n)] \quad (13.21)$$

The controller $K(s)$ is decomposed into USP ($Z(s)$) and controller $K_{\text{USP}}(s)$ so that $K(s) = K_{\text{USP}}(s)[I - K_{\text{USP}}(s)C(s)]^{-1}$. The augmented plant $\tilde{P}(s)$ with controller $K_{\text{USP}}(s)$ gives the same performance as delayed plant $G(s)$ with $K(s)$.

13.4 PARAMETERIZATION OF TWO-DOF CONTROLLERS

The 2-DOF controllers provide the advantage of complete separation of reference tracking and feedback properties. This is very advantageous to provide a good compromise between setpoint tracking and disturbance rejection if input and disturbance have a different behavior toward error.

13.4.1 DESIGN OF FEEDBACK CONTROLLER

The parameters of the feedback controller are found out by taking the performance criteria as minimization of H_2 norm of closed-loop transfer function between erroneous output $z(s)$ and disturbance input $w(s)$ as shown in Figure 13.1. The suboptimal H_∞ control problem is

$$\min\|T_{z\omega}\|_2 \le \gamma \quad (13.22)$$

where

$$T_{z\omega} = F_l(G,K) = F_l\left(\tilde{P}, K_{\text{USP}}\right) \quad (13.23)$$

F_l is the lower-fractional transform. Objective function (13.22) is minimized by tuning the parameters of the weighting function used for converting the system to lower-fractional transform (LFT). For LFT plants, the observer controller is found out using state feedback parameter S and state observer parameter L, which are chosen such that $A + B_2S$ and $A + LC_2$ have closed-loop poles LHS of s-plane. The standard H_2 problem for $G(s)$ is developed to find a stabilizing controller K, which minimizes the H_2 norm of the transfer function T_{zw} given by (13.22). The solution to this minimization problem involves the Riccati equation corresponding to the Hamiltonian matrices H and J, which are given as:

$$H = \begin{bmatrix} P_a & 0 \\ -P_{c_1}'P_{c_1} & -P_a' \end{bmatrix} - \begin{bmatrix} P_{b2} \\ -P_{c_1}'P_{d12} \end{bmatrix} \begin{bmatrix} P_{d12}'P_{c_1} & P_{b2}' \end{bmatrix} \quad (13.24)$$

$$J = \begin{bmatrix} P_a' & 0 \\ -P_{b1}P_{b1}' & -P_a \end{bmatrix} - \begin{bmatrix} P_{c2}' \\ -P_{b1}P_{d21}' \end{bmatrix} \begin{bmatrix} P_{d21}P_{b1}' & P_{c2} \end{bmatrix} \quad (13.25)$$

Here, the notations used for plant parameters are the same as in (13.19). (') notation represents transpose. For designing K, assume the following conditions:

i. (A, B_2) is stabilizable, and (C_2, A) is detectable.

ii. $\begin{bmatrix} A - j\omega I & B_2 \\ C_1 & D_{12} \end{bmatrix}$ has full column rank $\forall \omega \in R$.

iii. $\begin{bmatrix} A - j\omega I & B_1 \\ C_2 & D_{21} \end{bmatrix}$ has full row rank $\forall \omega \in R$.

iv. $D_{12}^* D_{12} = I$ and $D_{21} D_{21}^* = I$.

If the conditions (i)–(iv) hold, then there exists a unique optimal H_2 controller K for the plant $G(s)$ and is given by $K = K_{usp}(1 - ZK_{usp})^{-1}$,

$$K_{usp} = \begin{bmatrix} P_a + P_{b_2}S + LP_{c_2} & \vdots & -L \\ \cdots & \vdots & \cdots \\ S & \vdots & 0 \end{bmatrix} \quad (13.26)$$

where

$$S = -\left(P_{b_2}' X + P_{d_{12}}' P_{c_1} \right) \quad (13.27)$$

$$L = -\left(YP_{c_2} + P_{b_1}P_{d_{21}}' \right) \quad (13.28)$$

which are written in terms of augmented plant parameters and the stabilizing solution of Riccati equation associated with H and J. The Riccati equation solution can be obtained by MATLAB function "gcare" given as:

$$X = \text{gcare}(H) \quad (13.29)$$

$$Y = \text{gcare}(J) \quad (13.30)$$

13.4.2 DESIGN OF FEEDFORWARD CONTROLLER

After finding out feedback controller, the parameters of fixed structure feedforward controllers are found out by minimizing the integral of square of error of step response of transfer function $T_{yr}(s)$ (between output $y(s)$ and input $r(s)$) and reference model T_{ref}, as shown in Figure 13.3. The fixed structure polynomial MIMO controller is taken as

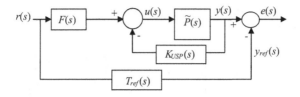

FIGURE 13.3 Two-DOF controller structure.

$$F(s) = \begin{bmatrix} \dfrac{f(1)}{s+f(2)} & \dfrac{f(3)}{s+f(4)} \\[2mm] \dfrac{f(5)}{s+f(6)} & \dfrac{f(7)}{s+f(8)} \end{bmatrix} \qquad (13.31)$$

So, the objective function to be minimized is

$$e(t) = \min_{f} \int_{0}^{t} \left| y(t) - y_{\text{ref}}(t) \right|^2 dt. \qquad (13.32)$$

13.5 SIMULATION EXAMPLE

Wood and Berry distillation column is taken for simulation to implement the design methodology and is defined by the transfer function matrices as

$$G(s) = \begin{bmatrix} \dfrac{12.8e^{-s}}{16.7s+1} & \dfrac{-18.9e^{-3s}}{21s+1} \\[3mm] \dfrac{6.6e^{-7s}}{10.9s+1} & \dfrac{-19.4e^{-3s}}{14.4s+1} \end{bmatrix} \qquad (13.33)$$

$$G_d(s) = \begin{bmatrix} \dfrac{3.8e^{-8.1s}}{14.9s+1} \\[3mm] \dfrac{4.9e^{-3.4s}}{13.2s+1} \end{bmatrix} \qquad (13.34)$$

The various variables in this system are as follows:

- **Output Variables**:
  - Distillate methanol (mol %).
  - Water (mol %).
- **Manipulated Variables**:
  - Reflux flow rate (lb/min).
  - Steam flow rate (lb/min).
- **Disturbance**:
  - Unmeasured flow rate.

For designing the USP, the delayed MIMO system is represented by delay-free part $P(s)$ and delayed matrix $\Gamma(s)$.

$$P(s) = \begin{bmatrix} \dfrac{12.8}{16.7s+1} & \dfrac{-18.9}{21s+1} \\ \dfrac{6.6}{10.9s+1} & \dfrac{-19.4}{14.4s+1} \end{bmatrix} \qquad (13.35)$$

$$\Gamma(s) = \begin{bmatrix} e^{-3s} & 0 \\ 0 & e^{-s} \end{bmatrix} \qquad (13.36)$$

The delayed matrix $\Gamma(s)$ is generated by taking the largest common delay corresponding to each input in "adj $G(s)$" [27]. The plant $P(s)$ is transformed to LFT using weighting function $W(s)$ whose parameters "$w_i(s)$" are tuned using GA by minimizing the objective function given in (13.22).

$$W(s) = \begin{bmatrix} \dfrac{w_1(s)}{s+w_2(s)} & 0 \\ 0 & \dfrac{w_3(s)}{s+w_4(s)} \end{bmatrix} \qquad (13.37)$$

The initialization parameters used for GA are the following:

Length of the string = 16 bits, population size = 20, crossover probability = 0.8, and mutation probability = 0.01.

The value of σ is taken as -0.07 for splitting the complex plane into two parts: stable and unstable.

Figure 13.4 shows the load rejection ability with minimum value of the objective function as 0.3779. The weighting function obtained using objective function (13.22) is

$$W(s) = \begin{bmatrix} \dfrac{0.5007}{s+0.5995} & 0 \\ 0 & \dfrac{0.3727}{s+0.4744} \end{bmatrix} \qquad (13.38)$$

The setpoint tracking response is shown in Figure 13.5, and the feedforward controller is obtained using objective function (13.32) and taking reference model as

$$T_{\text{ref}} = \begin{bmatrix} \dfrac{49}{(s+7)^2} & 0 \\ 0 & \dfrac{49}{(s+7)^2} \end{bmatrix} \qquad (13.39)$$

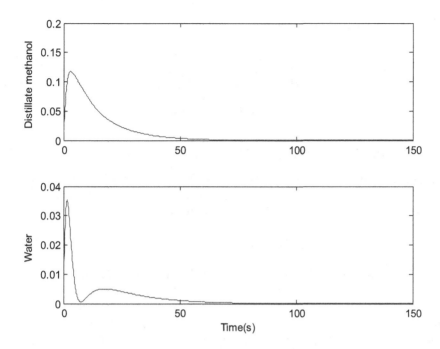

FIGURE 13.4   Disturbance rejection responses for impulse disturbance input.

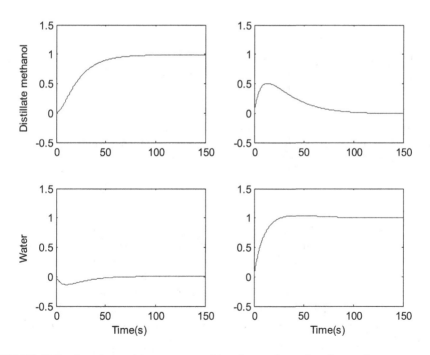

FIGURE 13.5   Setpoint tracking response with unit step change in reference input.

The feedforward controller obtained is given by

$$F(s) = \begin{bmatrix} \dfrac{-12.4155}{s+80.4319} & \dfrac{13.4734}{s+80.6026} \\ \dfrac{-4.7735}{s+80.0757} & \dfrac{8.9379}{s+80.9840} \end{bmatrix} \tag{13.40}$$

13.6 ROBUST STABILITY ANALYSIS

A good performance and robustness is achieved by the present design approach, which is proved by the spectral radius of $M\Delta$ structure. The necessary and sufficient condition for the system to be stable using this method is given as

$$\rho(M\Delta) \leq 1 \quad \forall \omega, \quad \forall \Delta, \quad \bar{\sigma} \leq 1 \tag{13.41}$$

where M is the weighting function multiplied by the system transfer function and Δ is the structured uncertainty. To demonstrate the robustness of the closed loop with the designed controller, multiplicative uncertainty Δ and uncertainty weighting function are considered. Δ is considered as the process actuator has 20% uncertainty at high frequency and 20% in the low frequency range. The weighting function considered has DC gain 0.3, crossover frequency 10, and high-frequency gain 20.

$$\Delta = \begin{bmatrix} \dfrac{s+0.2}{5s+1} & 0 \\ 0 & \dfrac{s+0.2}{5s+1} \end{bmatrix}; \quad W_u = \begin{bmatrix} \dfrac{20s+62.82}{s+209.4} & 0 \\ 0 & \dfrac{20s+62.82}{s+209.4} \end{bmatrix} \tag{13.42}$$

The spectral radius of the $M\Delta$ structure is plotted in Figure 13.6. It shows that for the assumed Δ, the magnitude is less than 1 over entire frequency range, which demonstrates the robust stability of the given system with uncertainty.

13.7 CONCLUSIONS

A design methodology has been developed for the MIMO systems with multiple time delays. A unified Smith predictor-based observer feedback controller has been designed to obtain disturbance rejection. For setpoint tracking, model reference fixed structure polynomial controller has been designed. The present 2-DOF controller design methodology is applied to Wood and Berry distillation column. The designed controllers have provided robust stability against multiplicative uncertainty with complete separation of setpoint tracking and disturbance rejection.

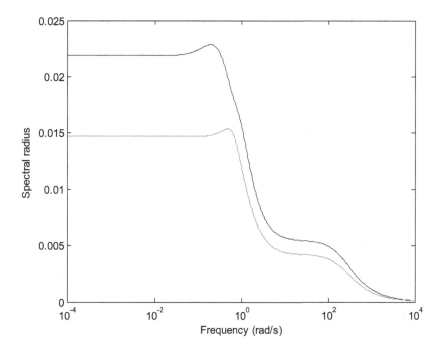

FIGURE 13.6   Spectral radius of perturbed system with multiplicative uncertainty.

REFERENCES

1. Q. C. Zhong, 2006. *Robust Control of Time-delay Systems*, Germany: Springer- Verlag London Limited.
2. A. T. Bahill, 1983. A simple adaptive Smith-predictor for controlling time-delay systems: A tutorial. *IEEE Control Syst. Mag.*, 3: 16–22.
3. F. Blanchini, et al., 2016. Stable LPV realisation of the Smith predictor. *Int. J. Syst. Sci.*, 47: 2393–2401.
4. K. Watanabe and M. Ito, 1981. A process model control for linear systems with delay. *IEEE Trans. Automat. Control*, 26: 1261–1269.
5. M. R. Matausek and A. D. Micic, 1996. A modified Smith predictor for controlling a process with an integrator and long dead time. *IEEE Trans. Automat. Control*, 41: 1199–1203.
6. M. R. Matausek and A. D. Micic, 1999. On the modified Smith predictor for controlling a process with an integrator and long dead time. *IEEE Trans. Automat. Control*, 44: 1603–1606.
7. S. Majhi and D. P. Atherton, 1999. Modified Smith predictor and controller for processes with time delay. *IEE Proc. - Control Theory Applicat.*, 146: 359–366.
8. L. Mirkin and Q. C. Zhong, 2003. 2DOF controller parameterization for systems with a single I/O delay. *IEEE Trans. Automat. Control*, 48: 1–5.
9. S. Alcantra, et al., 2009. General Smith predictors from an observer-controller perspective in *Proceedings of 18th IEEE International Conference on Control Applications and Intelligent Control*, St. Petersburg, Russia.
10. P. F. Beevi, et al., 2014. Enhanced IMC-PID controller design with lead-lag filter for unstable and integrating processes with time delay. *J. Chem. Eng. Commun.*, 201: 1468–1496.

11. P. F. Beevi, et al., 2016. Two degree of freedom controller design by AGTM/AGMP matching method for time delay systems. *Procedia Technol.*, 25: 20–27.

12. C. Yin, et al., 2014. Enhanced PID controllers design based on modified Smith predictor control for unstable process with time delay. *Math. Problems Eng.*, 2014: 1–7, article ID 521460.

13. H. Tian, et al., 2016. Predictor based two-degree-of-freedom control design for industrial stable processes with long input delay in *Proceedings of 35th Chinese Control Conference (CCC)*, Chengdu, China.

14. D. F. N. Rodriguez, et al., 2014. Observer based scheme for the control of high order systems with two unstable poles plus time delay. *Asia Pacific J. Chem. Eng.*, 9: 167–180.

15. P. Bistak and M. Huba, 2014. Model reference control of a two tank system in *Proceedings of 18th International Conference on System Theory, Control and Computing (ICSTCC)*, Sinaia, Romania.

16. M. Ajmeri and A. Ali, 2015. Two degree of freedom control scheme for unstable processes with small time delay. *ISA Trans.*, 56: 308–326.

17. N. Suwa and Y. Ishida, 2015. Controller design by frequency-domain fictitious reference iterative tuning using sliding DFT in *Proceedings of 3rd International Conference on Artificial Intelligence, Modelling and Simulation (AIMS)*, Kota Kinabalu, Malaysia.

18. F. Moya, et al., 2017. Pareto-based polynomial tuning rule for 2DoF PID controllers for time-delayed dominant processes with robustness consideration in *Proceedings of American Control Conference*, Seattle, WA, USA.

19. Q. C. Zhong, 2003. On standard H_∞ control of processes with a single delay. *IEEE Trans. Automat. Control*, 48: 1097–1103.

20. G. Tadmor, 2000. The standard H∞ problem in systems with a single input delay. *IEEE Trans. Automat. Control*, 45: 382–397.

21. L. Mirkin, et al., 2011. Dead-time compensation for systems with multiple I/O delays: A Loop-Shifting Approach. *IEEE Trans. Automat. Control*, 56: 2542–2554.

22. G. Meinsma and L. Mirkin, 2005. H_∞ control of systems with multiple I/O delays via decomposition to adobe problems. *IEEE Trans. Automat. Control*, 50: 199–211.

23. B. A. Ogunnaike and W. H. Ray, 1979. Multivariable controller design for linear systems having multiple time delays. *AIChE J.*, 25: 1043–1057.

24. Q. C. Zhong and G. Weiss, 2004. A unified smith predictor based on the spectral decomposition of the plant. *Int. J. Control*, 77: 1362–1371.

25. R. Majumder, et al., 2005. A unified smith predictor approach for power system damping control design using remote signals. *IEEE Trans. Control Syst. Technol.*, 13: 1063–1068.

26. A. S. Rao and M. Chidambaram, 2006. Decoupled smith predictor for multivariable nonsquare systems with multiple time delays. *J. Indian Inst. Sci.*, 86: 235–256.

27. C. Huang, et al., 2011. Design of decoupling smith control for multivariable system with time delays. *J. Central South Univ. Technol.*, 18: 473–478.

28. G. Galdos, et al., 2010. H_∞ controller design for spectral MIMO models by convex optimization. *J. Process Control*, 20: 1175–1182.

29. A. Nicoletti and A. Karimi, 2014. H∞ Smith Predictor design for time-delayed MIMO systems via convex optimization in *Proceedings of IEEE Conference on Control Application*, Antibes, France, pp. 1418–1424.

30. V. D. Oliveira, et al., 2015. *Robust Smith Predictor Design for Time-Delay Systems with H_∞ Performance*. No. EPFL-CHAPTER-213472, Switzerland: Springer International Publishing, pp. 287–307.

14 Renewable Energy Sources and Small Hydro Power Scenario in Mountainous Regions of Himalayas

Umesh C. Rathore
Govt. Hydro Engineering College

Sanjeev Singh
Maulana Azad National Institute of Technology (MANIT)

Pradeep Singh Thakur
RG Govt. Engineering College

Krishan Arora
Lovely Professional University

CONTENTS

DOI: 10.1201/9781003242277-14

14.1   INTRODUCTION

This chapter provides a brief outline of various types of renewable energy sources available and the need to harnessing these sources to meet the increasing energy demand to reduce gap between demand and supply and also to protect the environment from ill effects of pollutants generated from harnessing conventional sources of energy. Among different types of renewable energy sources, small hydro is the cleanest and best option of renewable energy source to cater the power demands of remotest part of the world where providing electrical power through conventional grid systems is a tedious task. Integrating small hydro with other sources of renewable energy such as wind and solar is best option to providing reliable power using smart grid in hybrid power generation system. This chapter also covers the small hydro scenario in mountainous terrains of Himalayas by studying the electrical loading patterns and rainfall trends in the recent past by collecting the data from one of the Himalayan states to emulate them for complete mountainous range and also to suggest the suitable energy conversion technology to harness this vast small hydro potential.

14.2   RENEWABLE ENERGY SOURCES

In the present energy scenario, harnessing renewable energy sources is the key to meet the increasing demand of energy. The ill effects on the environment caused by the harnessing of various conventional energy sources and consumption of energy have resulted in worldwide attention to find the alternative sources of energy with efficient and reliable energy conversion technology. However, an imbalance of energy consumption is still prevalent around the world mostly in developing and less developed countries across the world. The energy consumption is high in most of the already developed countries. On the other hand, the developing countries also need to use more energy to ensure economic growth. Oil and gas have already become too expensive, and with each passing day, their reserves are depleting. There have been few major energy crises in the world such as oil and energy crisis of the 1970s and also the 1990 oil price hike due to the Gulf War (Simoes et al., 2004). Energy is one of the major drivers of a growing economy in any country. Therefore, there is an urgent requirement of various renewable energy sources exploration, capacity enhancements, and simultaneously carrying out the desired energy reforms to deal with the energy deficit-related issues.

Due to the worldwide stress on harnessing renewable energy sources, many researchers have focused their research in this field and have proposed various methods and reliable technologies related to renewable energy conversion systems (Adhikari et al., 2003; Krishnadas & Ramachandra, 2010; Kathirvel & Porkumaran, 2011). The methods of harnessing energy from various energy sources are becoming important from the environmental conservation point of view and due to fast depletion rate of conventional sources of energy. The energy sources can be broadly classified into two categories known as conventional sources of energy and non-conventional sources of energy. The major portion of the world's electrical energy production is being produced by harnessing the traditional conventional sources of energy such as fossil fuels. Fossil fuels have limited reserves, and these sources will be finished totally in coming years. Also the harnessing of these conventional sources of energy leads to environmental degradation due to emission of harmful gases in the atmosphere, which pollutes the air and leads to environmental and health concerns all over the world. Therefore, people are forced and motivated to look for the alternative sources of energy that can replace conventional sources of energy to meet the ever-increasing demand of energy for the economic development and also to save the environment. The alternative sources of energy are also called non-conventional sources of energy or renewable sources of energy. These energy sources include wind, small hydro, solar, biomass, tidal, ocean thermal, and geothermal energies. These are called renewable sources of energy because these are continuously replenished by natural processes. In renewable energy system, the energy contained in sunlight, wind, water, sea waves, geothermal heat, and biomass is converted into heat or electricity. It is called renewable energy because most of the energy comes either directly or indirectly from the sun and wind and which can never be exhausted. About 20% global electricity generation comes from harnessing of these renewable energy resources (Simoes et al., 2004). With the passage of time, more and more emphasis is being given around the world to maximize the harnessing of renewable energy sources. The development of technologies in energy conversion processes has also led to the widespread use of renewable energy sources in producing electricity. Generally, the renewable energy technology is suited to rural and remote areas where energy is essential for the development. Globally, an estimated amount of more than 3.5 million households get power from small solar-based photovoltaic systems (Adhikari et al., 2003). Micro-hydro systems are providing energy to a large amount of world population living in remote hilly locations by feeding the local small grid network or in isolated mode (Singh & Kasal 2006; Rajagopal & Singh, 2011; Chilipi et al., 2014; Kalla et al., 2016). Similarly, vast wind energy potential is being harnessed at a very fast growth rate all over the world (Celik, 2003; Venkataraman et al., 2010). With the rapid advancement in technology related to induction generators, which are most suitable for the small hydro and wind energy conversion applications, the exploration and harnessing of these energy resources has grown up rapidly all over the world. The advancement in technology related to photovoltaic conversion and solar panels has led to widespread installation of large solar-based power plants around the world to generate electricity from solar energy. Solar energy can either be directly used in heating or can be converted into electricity by using photovoltaic principle or concentrated solar power technologies (Adhikari et al., 2003). Most of these renewable sources of energy can either be used to produce electrical energy in stand-alone mode or in hybrid mode. In hybrid mode, two or more types of renewable energy sources are used to generate electricity to feed electrical grid to increase its reliability in feeding power (Rathore, 2017).

14.3 HYBRID POWER GENERATION SYSTEM

When we combine two or more than two different power generation systems consisting of mixed renewable energy sources, then it is called hybrid power generation system (Andrade et al., 2009; Deb et al., 2012; Srivastava et al., 2012). In a hybrid generation system, there are two or more energy sources used to feed a grid, especially in remote areas. The motivation for hybrid power generation system is that the grid power is often unreliable, limited, or even non-existent in remote and rural areas. The main challenge of hybrid power generation system is to provide a reliable, low-cost power management system that is scalable. The combination of different renewable sources provides a good uninterrupted power system. Different renewable energy conversion generators would complement each other. However, while going for hybrid power generation system, it requires fulfilling certain criteria and certain factors such as location, time, and user's need for power have to be taken into consideration (Deb et al., 2012). Thereafter, the suitability of the type of renewable sources is decided. A hybrid power system incorporating various renewable energy sources offers greater reliability than any of the single energy source as power supply does not depend on a single source. Renewable energy sources are key in hybrid system as some of the main renewable sources such as wind and solar are seasonal (Deb et al., 2012) as these may not be available all the time. Similarly, in case of small hydro power plants, the water flow rate is not consistent. In such cases, the hybrid power generation plays the key role in maintaining the reliability of the power fed to a particular area. In the case of wind–hydro hybrid system, at the time of weak wind, the storage water capacity of hydropower plant can be utilized to meet the load demand by running hydropower plant to its full capacity. Similarly, at the time of high wind velocity, the full potential of wind energy can be harnessed and the wind power plant can share the majority of load demand at that time. This will relieve some load of hydropower plant, and at that time, water can be stored in case of storage system which can later be used when wind power plant is not running to its full capacity. Hence, the integration of two energy sources helps in increasing the output power of the system with increased reliability. A schematic representation of a small hydro–wind–solar and diesel engine generator-based hybrid power generation system is shown in Figure 14.1 (Rathore, 2017).

The advantage of the hybrid power system is that it improves the quality and availability of power. It also leads to a reduction in required generating capacity of individual systems such as solar and wind energies as the total load is shared between all these sources as per availability. Another combination of renewable energy sources in hybrid power generation consists of wind, solar, diesel generator, and power supply line. This helps to increase the system efficiency as well as to achieve a greater balance in power supply. By using the combination of these energy sources, a greater output can be obtained from turbines in winter and peak wind seasons and similarly during the summer, solar panels would produce the peak output. In a hybrid power generation system, the base load is covered by the largest available renewable sources that are firmly available in the system and other intermittent sources meet the peak load in a small electric grid (Lal et al., 2011). Thus, the optimization of the type of renewable energy sources suited for the particular

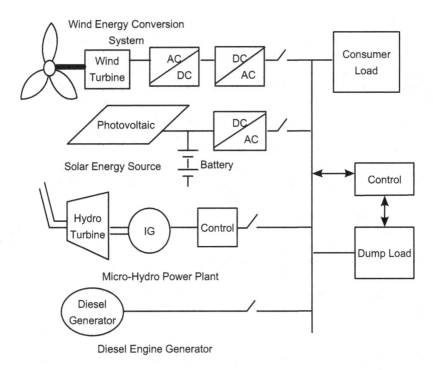

FIGURE 14.1   Pico-hydro, wind, solar, and diesel generator-based hybrid power generation system.

area has to be done for the reliability of the power system feeding isolated areas. Various software techniques are available to optimize the use of renewable energy sources for a given situation. One such software is the Hybrid Optimization Model for Electric Renewable (HOMER) for optimizing the renewable energy sources (Lal et al., 2011). Various types of renewable energy sources, energy storage systems, and their applicability in terms of cost and performance have been analyzed by researchers (Lal et al., 2011). The HOMER software is used to study and design the hybrid alternative energy power model. The optimum hybrid configuration is determined using this HOMER software (Lal et al., 2011). The use of HOMER software can be very useful in identifying the suitability of renewable energy sources in rural areas as in countries such as India the majority of the population (more than 70%) resides in these rural areas. Also, there is a need for greater reforms in power sector due to the increasing gap between power generation and the ever-increasing demand for electrical power. Under such cases, the role of renewable energy sources and their effective utilization along with other conventional energy sources become crucial. The generated power from the renewable energy sources is generally fluctuating in nature due to the environmental and seasonal changes. To optimize the harnessing of renewable energy sources effectively, there is a need to determine the proportions of various renewable energy sources on the basis of techno-economic feasibility. It is also necessary to consider the factors that influence the conventional energy sources.

One such factor is the diversion of peak demand to the power generation from renewable energy sources. The other factors include the introduction of energy storage systems, emission reduction plans, restructuring of energy policies, and emphasis of renewable energy sources. For developing the renewable energy sources, a cost–benefit analysis considering the capital cost, generation cost, and transmission losses needs to be carried out at the planning stage. Another analytical study on seasonal and spatial variation of solar irradiance for photovoltaic energy potential in Nepal Himalayas has also been explored by researchers to provide reliable power to people living in remote mountainous Himalayan regions (Adhikari et al., 2003).

Wind–solar hybrid is one of the combinations among various combinations in hybrid power systems. Various researchers have explored the possibility and performed an exploratory study of the feasibility of wind–solar hybrid power generation across various parts of the world (Krishnadas & Ramachandra 2010; Hirose & Matsuo, 2012). To enhance the reliability of wind–solar hybrid system, researchers have modeled the system using the artificial intelligence technique and better control systems. Due to the variable nature of the various renewable energy sources inputs, researchers have proposed various methods of energy storage such as batteries and ultra-capacitors to maintain the reliability of output power in a hybrid system (Goel et al., 2011; Srivastava et al., 2012).

Among the various combinations of various renewable energy sources in the hybrid power generation system, wind–hydro combination is most popular hybrid combination suitable for remote mountainous regions. Various researchers have explored and proposed the wind–hydro hybrid power generation system with various design and control features depending upon the availability of wind and water discharge at the input as well as at various loading conditions (Aktarujjaman et al., 2005; Goel et al., 2009; Dhanalakshmi & Palaniswami, 2011). The wind–hydro hybrid power generation system has numerous advantages, especially power reliability in remote locations feeding isolated loads. In small hydro power plants, the output power fluctuates comparatively less due to the less variation in water flow except in sudden and abrupt climate changes and during flash floods. Flow curves of small hydro power plants show seasonal variations, but these plants produce constant power except during the case when there is a severe drought period. A small hydro power plant can complement the wind power plants, and these are very useful in compensating for the fluctuating wind energy output. The wind–hydro combination in hybrid power generation system is most suitable for mountainous regions where apart from abundance hydropower potential, there is sufficient wind energy to support and complement the small hydro in that region. A study for the scope of wind and solar potential in Himachal Pradesh has been done to explore the possibility of wind and solar energy harnessing in remote regions to maintain the electrical power reliability when electricity supply through the conventional grid fails during harsh climatic conditions (Krishnadas & Ramachandra 2010). The wind–hydro hybrid power system is best suited to isolated load where load and frequency control is better achieved in the wind–hydro hybrid system. Another feature of the wind–hydro hybrid system is the improved reliability and an increase in the effective load-carrying capacity. The wind–hydro coordination is achieved by taking advantage of utilizing an extra portion of generating capability of the hydropower plants that are not considered dependable generating capacity due to water flow limitations on certain times. To maintain the power quality in hybrid generation systems, the study of dynamics of the wind–hydro system has also been done (Aktarujjaman et al., 2005). Suitable control mechanism is necessary in the hybrid system with energy storage equipment to maintain the reliability quality of output power under all operating conditions (Nasser et al., 2008; Nasser et al., 2009). Another main feature of the wind–hydro hybrid generation system is the production of electricity with zero emission. The methodology of control of distributed generation reported in the literature involves the use of conventional proportional–integral controller (Dhanalakshmi & Palaniswami, 2011) and power electronics-based controllers using voltage source converters for load frequency control (Goel et al., 2009). For effective coordination and better control in the wind–hydro hybrid system, the use of permanent magnet synchronous generator along with self-excited induction generator has also been reported in the literature (Goel et al., 2009). The squirrel cage induction generator is used in small hydro power plants along with variable speed permanent magnet synchronous generator for wind power generation in the hydro–wind hybrid system feeding isolated load. The reliability of the power system is important in hybrid power generation systems incorporating energy sources having intermittent nature of inputs such as wind energy. In distribution network, the effect of placement of distributed generators used in hybrid power generation system has been reported in the literature. The proper use of the distributed generation in distribution network helps in mitigating voltage sag and improving power quality of the system (Amanifar & Golshan, 2012).

14.4 SOLAR ENERGY

There is a history of thousands of years of harnessing solar energy for growing crops, for warming houses, and for drying various eatables and other things. Photovoltaic (PV) cells are made from silicon or other compatible materials that convert sunlight directly into electricity. Distributed solar systems generate electricity locally for homes through rooftop solar panels. Large solar farms can generate power for plenty of homes using solar concentrators. Solar energy systems do not harm the environment and also do not generate greenhouse gases.

The availability of solar insolation in Himalayan region is influenced by its topography, seasons, and also microclimate. The elevation gradient-based study shows that the tropical to wet-temperate zone receives higher GHI (>5 kWh/m^2/day) for a major part of the year compared to the higher dry-temperate to alpine zone (4–4.5 kWh/m^2/day) annually (Krishnadas & Ramachandra 2010). The comparatively energy-intensive livelihoods in tropical to wet-temperate zone could rely on the substantial solar energy available. Therefore, solar cookers in homes and communities should be encouraged to reduce the use of conventional fuel for cooking. The colder and higher elevation zones could utilize solar energy for room/water heating, which reduces dependence on wood-based fuel.

14.5 WIND ENERGY

Wind energy conversion technology has evolved quickly over the last 30 years with increasing rotor diameters and the use of latest power electronics-based devices to

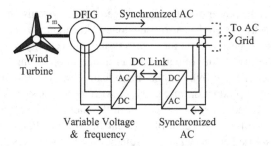

FIGURE 14.2   The schematic diagram of doubly fed induction generator used in WECS.

allow operation at variable speeds. Wind turbines produce electricity by using the wind force to further drive an electrical generator. Wind passes over the blades, generating lift and further producing a turning force. The rotating blades turn a shaft in a gearbox system. The gearbox increases the rotational speed to the value that is required by the generator, which further uses magnetic fields to convert the rotational energy into electrical energy. Generally, there are two types of wind turbines. One is fixed speed type, and the other is variable speed type. Fixed speed wind turbines are simple devices consisting of an aerodynamic rotor driving a low-speed shaft, a gearbox, a high-speed shaft, and an induction generator. Mostly, doubly fed induction generator is used. As the size of wind turbines has increased, the technology has moved from fixed speed to variable speed type turbines as there is a reduction in mechanical loads achieved with variable speed operation. The schematic representation of doubly fed induction generator used in wind energy conversion system is shown in Figure 14.2 (Rathore, 2017).

Wind resource availability in the complex terrain of Himalayan region has been explored considering the fact that wind speeds vary largely based on microclimatic conditions. Wind speed values mostly below 4 m/s showed that large-scale commercial power generation might not be feasible in these regions. The middle and higher elevation zones have relatively higher wind speeds compared to lower tropical zones (Krishnadas & Ramachandra, 2010). These speeds are favorable for small wind technologies such as agricultural water pumps, wind–photovoltaic hybrids, and space and water heaters, which might help in meeting part of the energy demand sustainably.

14.6   GEOTHERMAL ENERGY

There are very few locations in the entire Himalayan regions where one can find the geothermal source of energy, which is generally available inside the earth' core. This energy can be made available at the surface of earth either naturally in the form of hot springs or by drilling deep to bring the hot source in the form of steam or molten lava to earth surface, which can further be transformed to be used either directly or indirectly.

14.7   BIOMASS ENERGY

Biomass energy or bioenergy includes energy derived from the combustion or digestion of various organic materials such as wood, animal dung, and various agricultural

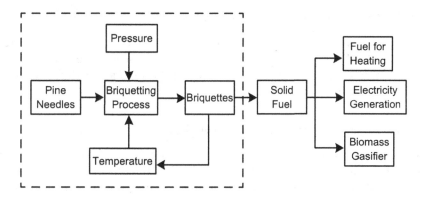

FIGURE 14.3 Biomass briquetting process using pine needles.

residues. In Himalayan regions, the agriculture residues and biogas-based bioenergy in higher-elevation zones are limited due to very less population. The other biomass resources available for power generation in mountainous regions are the pine needles. Pine needles are more prone to fire than other wastes such as leaves from other trees, paddy waste, and wood. These pine needles are highly susceptible to catch fire, and the whole forest finally ends up in ashes, resulting in loss of property. These pine needles can be used to generate electricity using briquetting process as shown in Figure 14.3. Collection of these pine needles through community projects will also eliminate the risk of forest fire in the hilly terrain (Rathore, 2017).

14.8   SMALL HYDRO POWER PLANTS

Small hydro power system is the most reliable and cost-effective renewable energy source. Micro- and pico-hydro power plants play a key role in meeting power requirements of remote, isolated hilly areas in isolated mode of operation by tapping hydro energy from small water streams, waterfalls, and canals of small water discharge. The captured energy from flowing water is converted into electricity. Small hydro projects are environment-friendly, have relatively less gestation periods, and require small capital investment as compared to large hydro power plants. These small hydro power plants play a significant role in meeting the electricity requirement of remote mountainous regions. The small hydro power system converts the potential energy of a mass of water flowing in a water stream with a certain head into electric energy at power station. The power generated is dependent mainly on water flow rate and head. The advantages of small hydro over other renewable energy sources are that these have a high efficiency (70%–90%) and a high capacity factor (>50%), compared with 10% for solar and 30% for wind energy conversion systems. Small hydro also has high level of predictability, varying with annual rainfall patterns taken from hydrology data and slow rate of change as the output power varies only gradually from time to time. Micro- and pico-hydro power plants are generally run-of-the-river type; therefore, run-of-the-river installations do not have the adverse effect on the environment as compared to large-scale hydropower plants (Rathore, 2017).

14.8.1   CLASSIFICATION OF SMALL HYDRO POWER PLANTS

Small hydro power plants are categorized in terms of their power output. Different countries have different norms, keeping the upper limit ranging from 5 to 50 MW. In India, small hydro power plants are classified by the Central Electricity Authority (CEA) as shown in Table 14.1.

14.8.2   ELEMENTS OF SMALL HYDRO POWER PLANTS

A hydro power system converts the potential energy of a mass of water into electric energy. In small hydro power plants, water is taken from the river by diverting it through an intake called weir. In case of medium- or high-head power plants, water is generally carried horizontally to the fore bay tank using a small canal. Before descending the water to turbine, it first passes through a settling tank or fore bay where suspended particulars are settled out. In the fore bay, rack of metal bars called trash rack are used to filter out debris. A pressure pipe called "penstock" carries the water from the fore bay to the hydro-turbine. The hydro-turbine is enclosed in the power house together with the electrical generator and other control equipment. After leaving the turbine, the water discharges down a "tailrace" canal back into the river or stream. Small hydro power plants are mainly run-of-the-river type with little or no water reservoir facility. In actual practice, the hydropower can be generated in mountainous locations where there are fast-flowing mountain water streams as shown in Figure 14.4 and in low land areas with wide rivers (Rathore, 2017).

14.8.3   POWER GENERATED IN SMALL HYDRO POWER PLANTS

The power in small hydro power plant is a function of water head (H) and the rate of flow of water called discharge (Q). The power theoretically available in a hydropower plant site is proportional to the difference in elevation between the source and the hydro-turbine, called the head H, and the water flow rate (Q) and is expressed as:

$$\text{Power} = \frac{\text{Energy}}{\text{Time}} = \frac{\text{Weight}}{\text{Volume}} \times \frac{\text{Volume}}{\text{Time}} \times \frac{\text{Energy}}{\text{Weight}} = \gamma QH \qquad (14.1)$$

The power (in watts) is $P = \gamma \cdot Q \cdot H_N$, where H_N is the net head (gross head – head loss). The power in kW from the nozzle is given as:

TABLE 14.1

Categories of Small Hydro Power Plants

S. No.	Type	Power Station Capacity	Generating Unit Rating
1	Pico	Up to 10 kW	Up to 10 kW
2	Micro	Up to 100 kW	Up to 100 kW
3	Mini	101–2,000 kW	101–1,000 kW
4	Small	2,001–15,000 kW	1,001–5,000 kW

FIGURE 14.4 Photographs of mountainous terrain suitable for installing SHP in the Himalayan region.

$$P_i(\text{kW}) = \frac{\gamma \times Q \times H_N}{1,000} = 9.81 \times Q \times H_N \qquad (14.2)$$

It shows that the power in the hydropower system depends on the net head and the flow rate of water. The higher the head, the better it is because there is a need of less flow rate to produce a given amount of power.

14.8.4 CONTROL REQUIREMENTS IN SMALL HYDRO POWER PLANTS

The main requirement of the consumers is to get the electrical power at rated frequency and voltage from the electrical system. Therefore, for maintaining these requisite parameters, certain controls are required in the power system at generation level. Usually, the system voltage is maintained by controlling the excitation voltage of the alternator and the desired frequency is maintained by controlling the speed of turbines according to the consumer load variation. To get desired frequency, it is necessary to maintain the generator shaft at a constant rotational speed. The speed of the small hydro power generator can be changed when loads are added or subtracted from the electrical system.

The frequency of the generated voltage and generator shaft speed are related as:

$$f = \frac{P \times N}{120} \qquad (14.3)$$

The system frequency can be maintained constant by removing the mismatch between electrical generator and consumer load. For this purpose, governor is used to control the speed of generator shaft as per the load requirement. Governor controls the water flow to the turbine as per the load requirement. Conventional governors are categorized as mechanical-hydraulic governors, electro-hydraulic governors, or mechanical governors. Mechanical-hydraulic governors are usually large devices that are used in large hydro power systems. These are costly and difficult to install, making their usage in small hydro power plants uneconomical. On the other hand, electro-hydraulic governors are complex devices and require accurate design and are also costly. They include complex guide vanes, inlet valves, and jet deflectors. Other types of governors used for controlling the water flow to the hydro-turbine include spear valve and servomotor along with PID controller. Spear valve is used for continuous flow control, and servomotor is used to operate the spear valve for flow control. Conventional governing systems are costly and complex and are not actually suited for small hydro power plants, which are isolated in nature. Therefore, load-side control is usually preferred in these small hydro power plants.

14.8.5 PICO-/MICRO-HYDRO POWER PLANTS AND THEIR CONTROL

In usual practice, the capacity of pico-/micro-hydro power plants is calculated on the basis of hydrology-based data which include water discharge, the amount of rainfall recorded in the previous few years, and the net water head available on the selected site. On the basis of recorded data, variation in water discharge has been observed in the past few years. This may be due to the effect of global warming on the overall environment causing erratic trends in rainfall in the entire mountainous region affecting the water discharge rate. This further affects the generated output of small hydro power plants installed in these regions. This water discharge variation has less effect on low-rated pico-hydro power plants with a rating of less than 100kW as these plants use uncontrolled hydro-turbines and have the continuous availability of desired water flow rate. These low-rated hydro-power plants are installed in mountainous regions feeding isolated load used by thinly populated villages having small electrical load. These small hydro power plants operating in isolated mode (Simoes et al., 2006) are best suited to these remote locations, which are usually difficult to connect with the conventional grid system. Apart from this, the daily power requirement varies during a day in these areas as most of the electrical load comprises of lighting and heating in nature. Due to the very low ratings, pico-hydro power plants are categorized as constant power-driven prime mover plants. This is because the water flow rate and head remain constant most of the time. In these types of isolated pico-hydro power plants, electronic load controllers (ELCs) are more suitable for control and they have resulted in increased reliability of the small hydropower system (Bonert et al., 1998; Singh et al., 2004; Bansal, 2005; Chatterjee et al., 2006; Murthy et al. 2006; Singh et al., 2010; Idjdarene et al. 2010; Karthikeyan et al. 2011; Kasal et al., 2011; Chilipi et al., 2014; Kalla et al., 2016; Rathore, 2017). An ELC is a power electronics-based device that is designed to control the output power of the small hydro power systems. It maintains almost constant load on the turbine and results in generating desired voltage and frequency. The controller controls the variation in the main consumer load by automatically changing the power dissipated in a resistive dump load so that the load seen by the generator is always constant. Water heaters are generally used as dump or ballast loads. Continuous feedback of frequency is taken for its operation. The frequency of the generated output is directly proportional to the speed of the turbine. The major advantages of ELC are that this control system has no moving parts and also this control system is more reliable and maintenance-free. The disadvantage of this control system is that it has no flow-controlling devices as generally used in the conventional governor control system. Therefore, there is a wastage of precious energy which could have otherwise been used in some other applications. However, the power lost in resistive heaters can be used for some heating purposes to avoid the wastage of power (Rathore, 2017).

14.8.6 WATER DISCHARGE AND ELECTRICAL LOADING PATTERN IN REMOTE MOUNTAINOUS REGIONS OF HIMALAYAS

The entire Himalayan region has plenty of natural water sources suitable for installing large and small hydro power plants. Mostly, the pico- and micro-hydro power plants are constructed in mountainous terrain of Himalayan states. Due to the disturbances in overall environment around the world and inconsistent rainfall, the output power generated in these power plants is not consistent. Figure 14.5 shows the monthly average water discharge recorded for 2 years in 1 MW micro-hydro power plant located in the Himalayan region (Rathore, 2017).

The very similar conditions exist in the other mountainous regions, and the water discharge rate depends upon the average rainfall and amount of precipitation of snow in the peaks in the concerned catchment area which feeds these hydropower plants. The recorded data as shown in Figure 14.6 show the annual power generation for 2 years. It shows that the generation of power varies due to the variation in water discharge rate. The generated power is maximum during rainy season in the months from July to September and also during the first 3 months of the year when there is sufficient rainfall due to western disturbances in mountainous Himalayan regions (Rathore et al., 2014; Rathore, 2017).

Also, the load requirement in these remote mountainous locations varies as per the daily power requirement. The electrical load mostly comprises of lighting and heating in the villages. The recorded average data of daily electrical load requirement for a period of 1 year from one of the remote Himalayan regions are as shown in Figure 14.7 (Rathore, 2017).

The nature of the electrical load in remote villages is generally lighting and heating load. Apart from this, there is no industrial load in these small hamlets, except

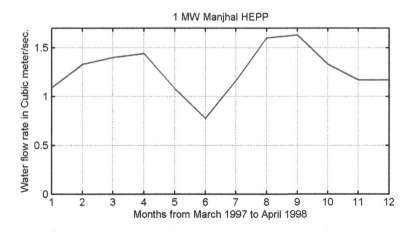

FIGURE 14.5 Monthly water discharge rate.

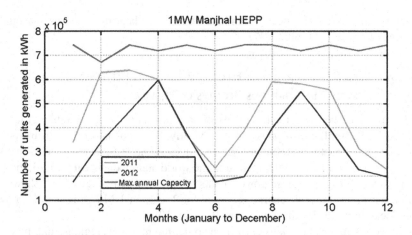

FIGURE 14.6 Annual kWh generations in 1 MW small hydro power plant.

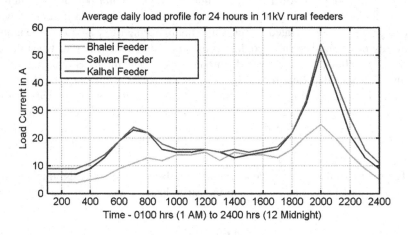

FIGURE 14.7 Electrical loading pattern in remote rural areas in hilly regions.

for very few flour mills that are mostly operated for a small period of times. The recorded data show the maximum load during the peak load hours in the evening time and for a small period in the morning hours. While for the rest of the time, especially in the midnight hours, the consumer load is very less ($\leq 20\%$ of the maximum load) on these power plants in isolated mode. Therefore, this kind of loading pattern is crucial as a special control mechanism is required for generating power at desired voltage and frequency under varying load conditions in isolated mode of operation. The use of induction generators and suitable ELC would be useful due to their adaptability to load changes and keeping the generated voltage and frequency under desired limits (Simoes et al., 2006). Usually, the frequency in self-excited induction generator (SEIG) is maintained under desired limits by using resistive dump load so that the load seen by generator is always constant. Figure 14.8 shows the schematic

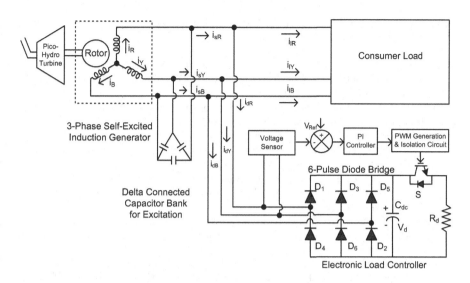

FIGURE 14.8   Use of ELC in isolated pico-hydro power generation system using SEIG.

representation of the use of ELC in isolated pico-/micro-hydro power generation systems (Rathore et al., 2014, Rathore, 2017).

Since resistive-type heaters are used in the dump load, there is a wastage of useful power during the off-load period. To avoid this wastage, we can use this energy in other heating applications, in water pumps in lift water schemes, or in heating the domestic houses in cold environments (Rathore, 2017).

14.9   CONCLUSIONS

In this chapter, a brief note on various types of renewable energy sources and their need and harnessing techniques have been presented. The concept of small hydro power plants, which is a type of renewable energy source, has been covered, and various data related to small hydro power, which were recorded in the Himalayan region, have been presented. The analysis of the data is very useful in exploring and selecting the type of renewable energy sources harnessing technology that will be useful to meet the electrical power requirements of people residing in these remote regions in the Himalayas.

REFERENCES

Adhikari, S., Adhikary, S., Umeno, M. (2003). PV energy potential in Nepal Himalayas: analytical study on seasonal and spatial variation of solar irradiance for PV. *Proc. 3rd World Conference on Photovoltaic Energy Conversion*, 11–18 May, 2003, Osaka, Japan, pp. 2027–2030.

Aktarujjaman, M., Kashem, M. A., Negnevitsky, M., Ledwich, G. (2005). Dynamics of a hydro-wind hybrid isolated power system. *Proc. Australasian Universities Power Engineering Conference (AUPEC 2005)*, 25–28 Sept. Hobart, Australia, pp. 231–236.

Amanifar, O., Golshan, M. E. H. (2012). The effect of placement of distributed generation on mitigating voltage sag in distribution network. *Proc. IEEE EPDC*, Canada, 2–3 May, pp. 1–6.

Balvet, V., Maubois, J.-E. (2009). Massive data acquisition for the short term wind power forecasting in a large grid: the hydro-Quebec SAGIPE system experience. *Proc. IEEE PSCE'09*, 15–18 Mar., pp. 1–7.

Bansal, R. C. (2005). Three-phase self-excited induction generators: an overview. *IEEE Transactions on Energy Conversion*. 20(2), 292–299.

Bonert, R., Rajakaruna, S. (1998). Self-excited induction generator with excellent voltage and frequency control. *IEE Proceedings - Generation, Transmission and Distribution*. 145(1), 33–39.

Celik, A. N. (2003). Energy output estimation for small- scale wind power generators using weibull-representative wind data. *Journal of Wind Engineering and Industrial Aerodynamics*. 91, 693–707.

Chatterjee, J. K., Perumal, B. V., Gopu, N. R. (2006). Analysis of operation of a self-excited induction generator with generalized impedance controller. *IEEE Transactions on Energy Conversion*. 22(2), 307–315.

Chilipi, R. R., Singh, B., Murthy, S. S., Madishetti, S., Bhuvaneswari, G. (2014). Design and implementation of dynamic electronic load controller for three-phase self-excited induction generator in remote small-hydro power generation. *IET Renewable Power Generation*. 8(3), 269–280.

Deb, G., Paul, R., Das, S. (2012). Hybrid power generation system. *International Journal of Computer and Electrical Engineering*. 4(2), 141–144.

Dhanalakshmi, R., Palaniswami, S. (2011). Load frequency control of wind diesel hydro hybrid power system using conventional PI controller. *European Journal of Scientific Research*. 60(4), 612–623.

Goel, P. K., Singh, B., Murthy, S.S., Kishore, N. (2009). Autonomous hybrid system using SCIG for hydro power generation and variable speed PMSG for wind power generation. *Proc. IEEE PEDS*, Canada, pp. 55–60.

Idjdarene, K., Rekioua, D., Rekioua, T., Tounzi, A. (2010). Performance of an isolated induction generator under unbalanced loads. *IEEE Transactions on Energy Conversion*. 25(2), 303–311.

Kalla, U. K., Singh, B., Murthy, S. S. (2016). Modified electronic load controller for constant frequency operation with voltage regulation of small hydro-driven single phase SEIG. *IEEE Transactions on Industry Applications*. 52(4), 2789–2800.

Karthikeyan, A., Nagamani, C., Ilango, G. S., Sreenivasulu, A. (2011). Hybrid, open loop excitation system for a wind turbine-driven stand-alone induction generator. *IET Renewable Power Generation*. 5(2), 184–193.

Kasal, G. K., Singh, B. (2011). Voltage and frequency controllers for asynchronous generator-based isolated wind energy conversion system. *IEEE Transactions on Energy Conversions*. 26(2), 402–416.

Kathirvel, C., Porkumaran, K. (2011). Technologies for tapping renewable energy: a survey. *European Journal of Scientific Research*, 67(1), 112–118.

Krishnadas, G., Ramachandra T.V. (2010). Scope for renewable energy in Himachal Pradesh, India – a study of solar and wind resource potential. *Proc. Lake-2010: Wetlands, Biodiversity and Climate Change*, Canada, pp. 22–24

Lal, D. K., Dash, B. B., Akella, A. K. (2011). Optimization of photo voltaic/wind/micro hydro/diesel hybrid power system in HOMER for study area. *International Journal on Electrical Engineering and Informatics*, 3(3), 307–325.

Murthy, S. S., Singh, B., Gupta, S., Kulkarni, A., Sivarajan, R. (2006). Water, water... anywhere-field experience on a pico-hydel system using self excited induction generator and electronic load controller. *IEEE Industry Applications Magazine*. 12(4), 65–76.

Rajagopal, V., Singh, B. (2011). Improved electronic load controller for off-grid induction generator in small hydro power generation. *Proc. IEEE, IICPE' 2010*, Canada, pp. 1–7.

Rathore, U. C., Singh, S. (2014). Performance evaluation of 3-phase self-excited induction generator for remote mountainous region of Himalayas. *Proc. IEEE International Conference on Control, Instrumentation, Energy & Communication (CIEC)*, Jan. 2014, University of Calcutta, India, pp. 421–425.

Rathore, U. C. (2017). Design and Development of Voltage and Frequency Controller for Micro/Pico Hydro Power Generation System. *PhD Thesis*, Department of Electrical & Instrumentation Engineering, SLIET University, Longowal, Punjab, India.

Simoes, M. G., Farret, F. A. (2004). *Renewable Energy Systems – Design and Analysis with Induction Generators*. CRC Press, New York.

Singh, B., Murthy, S. S., Gupta, S. (2004). Analysis and design of STATCOM-based voltage regulator for self-excited induction generators. *IEEE Transactions on Energy Conversion*. 19(4), 783–790.

Singh, B., Rajagopal, V., Chandra, A., Al-Haddad, K. (2010). Development of electronic load controller for IAG based standalone hydro power generation. *Proc. of 2010 Annual IEEE India Conference (INDICON)*, Calcutta, India, 17–19 Dec., 1–4.

Srivastava, A. K., Kumar, A. A., Schulz, N. N. (2012). Impact of distributed generations with energy storage devices on the electric grid. *IEEE Systems Journal*. 6(1), 110–117.

Venkataraman, S. V., Iniyan, S., Suganthi, L., Goic, R. (2010). Wind energy potential estimation in India. *Proc. Clean Technology Conference & Expo*, Jun. 21–24, Anaheim, CA, 179–182.

15 A Comprehensive Review on Energy Storage Systems

A. Gayathri
Sri Krishna College of Technology

V. Rukkumani
Sri Ramakrishna Engineering College

V. Manimegalai
Sri Krishna College of Technology

P. Pandiyan
KPR Institute of Engineering and Technology

CONTENTS

DOI: 10.1201/9781003242277-15

15.1 INTRODUCTION

Energy systems are critical for harvesting energy from a variety of sources and transforming it into another form of energy, which is required for use in a variety of applications. Energy is used in various sectors, including utilities, industrial areas, home and office buildings, and transportation. Conventional energy provides energy based on the demand raised by the consumer so that such energy can be stored for later use. Non-conventional energies, on the other hand, are obtained when it is feasible and must be stored until they are required [1]. As of now, electrical energy is consumed as soon as it is generated, so electricity should always meet its demand based on the sectors it is consumed. The voltage and frequency stability of the power system may be affected by supply and energy demand imbalances. As a known fact, electricity-generating plants are located far away from the distribution sectors [2], so the cost of transmission lines based on the distance will also increase and losses will occur.

Every day, service companies and grid operators face the huge challenge of producing adequate energy to fulfill the unpredictable energy demands on the electric grid. The demand is high during the daytime because the greatest number of electrical appliances and machines are drawing power from the grid. As day turns to night, the world's energy consumption decreases, allowing the grid to get ready for future spikes in energy consumption.

In order to avoid the above-mentioned problems, both short-term and long-term storage systems are useful. Energy can be stored, and it can be portable based on the interests of the consumer sector. It is not easy to store electrical energy, and storing it can affect energy efficiency. It is imperative to have knowledge of the variations in current and the various materials that are being used for electricity storage [3].

Storing electrical energy can provide numerous benefits to energy systems, including the increased use of renewable energy sources, and good economic fulfillment also results in peak shaving and load balancing, damping oscillations, regulated frequency, power quality improvement, and better accuracy [1]. Energy storage can improve grid stability and overall system efficiency while also avoiding the use of fossil fuels and energy degradation, resulting in a greener environment [4].

In recent decades, various ESS have been used to collect and store energy for use when needed [5,6]. There are a variety of ESS, viz. electrochemical, chemical, electrical, thermal, mechanical, electrostatic, and electromagnetic energy storage methods. While each storage system has its benefits and drawbacks, a collective approach could benefit from a wider range of strategies.

This chapter examines different forms of ESS, focusing on their working principles as well as technological aspects. Additionally, an evaluation of various energy storage options is given in detail through an in-depth analysis of the application scenarios in Section 15.3. Section 15.4 provides a technical comparison of various energy storage technologies, which include new forms of energy storage devices and recent advancements. The challenges and issues of deploying energy storage devices are also presented in Section 15.5. In order to familiarize an energy storage system in future for all applications, brief research required in this field as discussed in Section 15.6. Finally, conclusions are drawn.

15.2 CLASSIFICATION OF ESS

Energy storage has been used for decades. It has been continuously improved to achieve the current level of development, and many of the storage types have reached a high level of matured systems. Due to their increased popularity, various storage categories have emerged. Based on the energy used in a definite form, ESS are categorized. For example, the storage of energy in electrochemical systems is based on the characteristics of specific energy and specific power capacity. This characteristic is given by "Ragone plot" [7], which helps to find out the capability of all storage types and compare them with different applications which require various energy storage options and on-demand rates of extraction of energy. The Ragone plot (Figure 15.1) can help you choose the best energy storage technology for your needs.

ESS can also be categorized according to their storage period. The "short-term energy storage" refers to storage that can store energy for a few hours to days, whereas the "long-term energy storage" refers to storage that can store energy for several weeks to 3–6 months. Examples of long-term storage include thermal energy storage, which can retain heat on the ground surface during the summer and the stored heat can be used as thermal energy in the winter. The ESS are classified as shown in Figure 15.2 based on their pattern and materials used [8,9]. A brief explanation for the above storage types is given in this section.

15.2.1 MECHANICAL ENERGY STORAGE (MES)

MES systems store input energy by taking advantage of gravitational and kinetic forces. Simple mechanical systems are commonly available to harness these forces

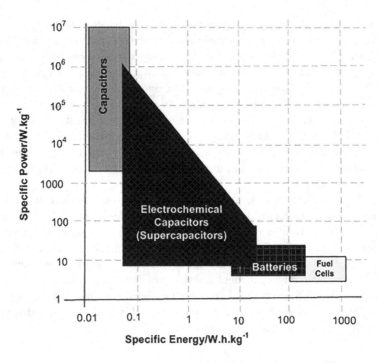

FIGURE 15.1 Ragone plot.

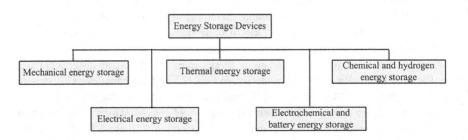

FIGURE 15.2 Classification of energy storage devices.

(for example, rotating a flywheel or lifting weights on a hill), but in order to use these simple forces efficiently and effectively, they require intricate technologies. These systems are made viable in real-world applications by the state-of-the-art computer control systems, innovative design, and high-tech materials. MES is widely employed for energy generation around the world [10]. This sort of energy storage is divided into three categories, as shown in Figure 15.3.

15.2.1.1 Pumped Hydroelectric Energy Storage (PHES)

The most frequently used mechanical energy storage system in hydropower plants is the pumped hydroelectric energy storage (PHES) method. For electricity generation, one of the most preferred sources of energy is hydroelectricity due to its success and

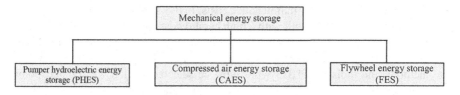

FIGURE 15.3 Types of mechanical energy storage.

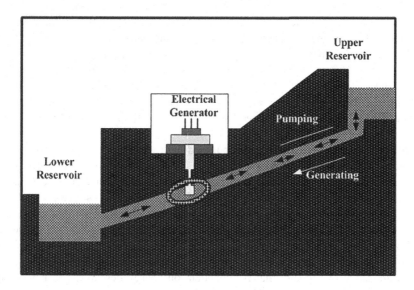

FIGURE 15.4 Pumped hydroelectric energy storage plant.

widespread use, as well as its long and successful track record of being both reliable and safe. Two reservoirs form the hydroelectric power plant: one in the lower valley and one in the mountains, with a higher level than the other. This can occur naturally in some locations, whereas in others, one or both reservoirs of water have artificially been created [10].

In full reservoir conditions, the reservoir performs identically to that of a hydroelectric plant with a high-head dam. When the demand for electricity increases, the upper reservoir's stored water is released, transforming the kinetic energy provided by its discharge into electrical power via rivers, pipes, and turbines. Although the water from the plant does not flow away into a downstream river, it is held in a lower reservoir for later use rather than returning to the river, as shown in Figure 15.4 [11]. Once the power generation phase is through, the same water that was conserved in the lower reservoir is pumped back uphill to the higher reservoir for storage (the same water that was saved in the lower reservoir) in pumping mode (Figure 15.4). More than 99% of the world's electrical storage capacity is provided by PHES, which has committed to providing 3% of the additional capacity for electricity generation.

The efficiency of PHES ranges from 70% to 80%, and the size of this type varies from 1,000 to 1,500 MW [12]. PHES has a long life expectancy of 50–100 years and is simple to operate with low maintenance costs. PHES has some drawbacks, including the fact that it is a large unit with a high capital cost and topographic limitations. In order to avoid such disadvantages, the feasible option of using underground PHES systems, otherwise called underground reservoirs, can be used [13]. The appropriate methods and tools for PHES types must be selected, which will be acceptable for commercial and social content [12]. To reach higher flexibility of operation and efficiency than traditional PHES systems, recent developments in PHES, such as variable-speed pumped hydroelectric storage (PHS) technology and by-pass pump or turbine arrangement, have been introduced to improve the operation hours.

Among these, variable-speed PHS technology acquires high capital costs, but provides a higher line of operation and performance efficiency when compared to traditional PHES types. The pumping capacity in small-capacity systems is designed to range from 60% to 100% and generates approximately 20% of full capacity [14]. When the efficiency is affected while using a single machine, there are different options to utilize multiple machines with different configurations. The definition of various control methods interrelated to different levels of operation of PHES systems [15] has been given.

Power converters are also used in pumped hydroelectric systems in both pumping mode and generating modes rather than mechanically controlled systems, and fixed-speed systems that can be used in microgrid applications. PHS development entails new development activities with various approaches to pumped hydroelectric systems [16]. The use of PHES in connecting intermittent renewable energy sources such as solar and wind has expanded in response to rising demand for ESS. The new design addresses the drawbacks of traditional PHES systems. Recent developments include a new design that eliminates tall water tank towers and long pipes, resulting in a wide range of capacity that is dependent on energy demand. The new design results in a stable pressure and fast discharge rate, which allows quick response to rapid demand variations [17].

15.2.1.2 Compressed-Air Energy Storage (CAES)

Compressed air with natural gas, which expands readily, results in the formation of modified gas, which is fed into a gas turbine connected to an electric generator to generate electricity [18]. This type is particularly well suited for high-capacity systems [19]. CAES is a method of storing energy generated at a particular point in time for later consumption. Utilities should be able to use excess energy from larger-scale power plants to fulfill the increased demand.

The above CAES plants have applications similar to PHES plants. However, the air or any other gases in a CAES plant are compressed and stored under pressure in an underground cavern or container rather than water pumping down from a lower to an upper pool during excess power periods (Figure 15.5).

When electrical energy is needed, compressed air is heated and expanded in an expansion turbine, which then drives a generator. In CAES, pressurized air is kept in an underground pothole or else an unused mine during the availability of excess energy. When there is an energy requirement, the compressed air that has been kept is released and taken out, which is fed to a turbine to produce electrical energy. The pothole is formed by drilling the slat or rock formation into the underground earth

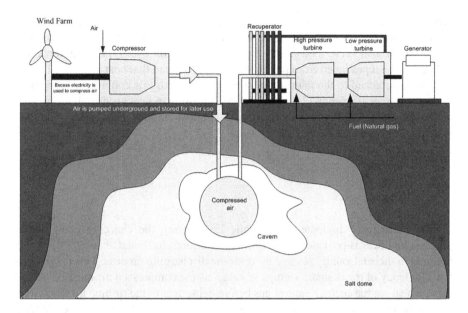

FIGURE 15.5   Compressed-air energy storage plant.

surface, or an already existing pothole is used. But such geological formations of potholes will not be available in all areas, so large steel tanks are installed under the ground surface, which maintain a high pressure. This sort of storage is economically attractive due to its potential to shift energy consumption times and the requirement to balance the grid's effects of alternative renewable energy integration [20].

Another option is to use the available energy to store dissolved air in appropriately insulated low-pressure storage tanks at ambient temperature. In comparison with pressurized air, this dissolved air has a lower loss rate, allowing it to be held at a moderate pressure. As a result, for long-term use, liquid air will be preferable than compressed air. When dissolved air is kept in small storage tanks, it is denser than when it is in the air. One survey says that 5,000 m³ of dissolved air is equal to 310,000 m³. And liquefied air gives higher efficiency than CAES [21]. It is possible to produce dissolved air using the Linde–Hampson cycle approach if there is sufficient energy available. Later, the Joule–Thomson effect was utilized for the expansion of the system. It also includes cryogenic turbines for expansion of valves, resulting in low-pressure operations, and it provides more dissolved air production rates with high efficiency [22]. Liquefied air is pumped at a high pressure during discharge. The high-pressure air is created by evaporating and then heating. Heat is produced by high-temperature mediums, such as gases produced by natural gas combustion. An increase in temperature enhances the work output and system's performance. Additionally, an increase in temperature can be obtained by using the air directly for the combustion process.

When liquid air evaporates during different cycles, it generates waste cooling power, generates additional work, and improves the system's overall efficiency to 80% [23]. The Brayton cycle or Rankine cycle uses cool air as a heat sink. When compressed air before it is converted into liquefied air, it produces waste heat, which is stored and utilized again to reheat the compressed air when it is fed into the turbines, which also serve as a heat source in the Brayton cycle. The Brayton cycle utilizes the heat produced from the liquefaction of air, and that heat is released to the evaporator attached to the natural gas storage system, so this coupling arrangement results in higher efficiency [24]. Based on the handling of heat produced during heat discharge and before the heat intake, CAES is classified into three types of storage systems. They are as follows:

1. Isothermal
2. Diabatic
3. Adiabatic.

Thermal energy is dissipated in diabatic ESS during the charging compression process, and heat is provided back while discharging. In the adiabatic type, heat is retained in thermal energy storage used for the discharging process. This results in a low efficiency of the diabatic storage systems, as the compressed air must be heated using a heat source such as natural gas before being sent to the turbine for discharge. In isothermal systems, the temperature is kept very close to the environmental temperature during compression and expansion, producing the necessary power for low compression and generating power when the expansion rate is high. So, this compression and expansion process can be handled by using piston machines [1].

15.2.1.3 Flywheel Energy Storage (FES)

A flywheel is a rotating mechanical system that can store rotational energy in the moment. Essentially, a flywheel is a rotating mass at the center of which a motor is used to propel the mass in order to generate energy when it is required. A turbine-like device produces energy using the spinning force, reducing the rate of rotation. The motor is used to replenish a flywheel by increasing its rotating speed once more. A flywheel may collect energy from intermittent energy sources and deliver consistent electricity to the grid over time. Additionally, flywheels are capable of instantaneous response to grid signals, allowing for frequency management and improved power quality [1].

Figure 15.6 depicts a FES system that stores energy with very low frictional losses by utilizing kinetic energy stored in a spinning mass. Through the use of coupling of motor–generator, electric energy is used to accelerate the mass to speed. The energy is discharged by removing the kinetic energy from a specific point in time using the same motor–generator. The amount of energy stored is proportional to the moment of inertia times the angular velocity squared of the object. To maximize the energy-to-mass ratio, the flywheel must spin at the quickest possible speed. Rapidly rotating objects are subjected to significant centrifugal forces. Dense materials can store more energy than low-density materials, but they are also exposed to increased centrifugal force, and as a result, they may be more prone to failure at lower rotational speeds.

The energy obtained from the flywheel is stored in the form of mechanical energy in order to facilitate the operation of electrical machines. This results in high power and higher energy density than is possible with other energy storage methods. The electric

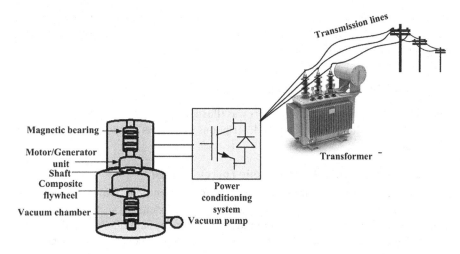

FIGURE 15.6   Flywheel energy storage.

machine acts as a generator or a motor when kinetic energy is transferred into and out of the flywheel, depending on the charging and discharging mode set by the user. Machines with permanent magnets are the most typically employed in flywheels because they give great efficiency, high power density, and minimal rotor side losses [25].

Based on the limitations of parameters such as speed, losses, vibration, cost, and noise, other machines such as induction machines, bearing less machines and reluctance machines, can be used in flywheels. When the flywheel rotates, it produces input energy which is stored as kinetic energy. Later, the stored energy is released and used as electrical energy when there is a demand for it. In order to remove frictional losses when the energy is stored for long-term use, the flywheel operates in a vacuum and ensures safety [26]. There is no need for lubrication requirements when we use magnetic bearings since they have no frictional losses. In high-speed applications, superconducting magnetic bearings can be used, but an additional cooling system is required in such models [27]. While using composite materials, a high rotational speed with higher energy density can be achieved when compared to batteries.

- **Advantages**
  1. High energy density
  2. More number of life cycles up to thousands of cycles
  3. The great operational life of the machine
  4. High efficiency
  5. Less impact on the environment.

- **Disadvantages**
  1. High cost
  2. Noise
  3. Difficult to maintain
  4. Safety concerns.

To increase the power density, a new design has been proposed, in which supercapacitors are used inside the rotating disk of a flywheel, which allows pulse power exchange along with a large storage capacity [28].

15.2.2 Electrical Energy Storage (EES)

There are numerous ways to store energy as illustrated in Figure 15.7 by the electrical energy storage system. The detailed descriptions of various forms of electric energy storage are as follows.

15.2.2.1 Superconducting Magnetic Energy Storage

Superconducting materials are known for conducting electricity with zero resistance. The various superconducting materials include chemical elements such as mercury and lead; alloys such as niobium; and ceramics such as magnesium diboride and lutetium. These elements are responsible for exhibiting their corresponding superconducting properties.

This storage system is extremely favorable in the context of energy storage. Faraday's law of electromagnetic induction states that a change in the flow of direct current into the superconducting coil induces a magnetic field around the coil, and thus, the generated energy is cryogenically cooled by liquid helium at 2.7 K or nitrogen vessels at −900°C, both of which are below the critical temperature of the superconducting coil [10].

When it comes to energy storage, this requires a huge number of coils to perform the storage function. A range of figures of coils, such as coil size, ability, and operating temperature, are meant to be considered for superconducting storage. The entire system is designed into various parts using the final configured factor values, as shown in Figure 15.8 [29].

1. Superconducting magnet and coil structure components
2. Refrigeration and protection system (cryogenic cooling)
3. Power conditioning system (power conversion: AC–DC or DC–AC)
4. Control system (demand monitoring).

The cost of the system is determined based on the components indicated above. With respect to the kind of application concerned, the system is being developed. Low-temperature superconducting devices are commonly used because, in addition to the quantity of power storage, the temperature range is also determined

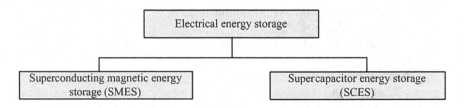

FIGURE 15.7 Types of electrical energy storage.

where power and temperature are directly related, so it is much more preferable for low-power storage applications in order to diminish the expensive cost.

This system is suitable for a variety of applications, including load leveling, power system regulation, power quality improvement, and uninterruptible power supplies. In future, self-responsive systems for higher-power applications will be developed, as well as superconducting coils that can conduct at room temperature.

• **Advantages**
    1. Superconducting magnetic energy storage (SMES) accounts for minimal losses when compared to other storage systems. As they carry away their conductive properties, there is no resistance, resulting in almost no loss.
    2. SMES have indefinite charge and discharge cycles. They maintain an efficiency of 95% of charge and discharge cycles.
    3. The discharge duration period of energy is faster. As per demand, the corresponding output can be obtained.

• **Disadvantages**
    1. The main requirements of the system are superconducting coils and a cooling system, as the materials are quite expensive, and there is some concern about the cost of implementing the system.

15.2.2.2 Supercapacitor Energy Storage

Generally, capacitors are known for storing charges when there is applied potential across the plates of a capacitor where charges get aligned toward the plates of the capacitor. The supercapacitor works on the same principle, but differs in the configuration structure of the components.

The supercapacitor is also called an ultra- or double-layer capacitor [30]. The varied structure of the supercapacitor consists of a charge collector of a polymer membrane and a carbon electrode soaked in the electrolyte, as shown in Figure 15.9. As the surface area of the electrode increases, charge accumulates on the electrode, increasing the capacitance of the capacitor's plates [31].

When a potential is applied across the capacitor, charges accumulate on the plate, forming Helmholtz double layer on the electrolyte's separator membrane. On the basis of their operating principle, supercapacitors can be divided into three categories as shown in Figure 15.10.

• **Advantages**
    1. The supercapacitor can store charges with a high energy density because its capacitance is 100 times greater than that of a normal capacitor.
    2. It can handle fast charging and discharging cycles.
    3. It has a greater efficiency of storage and long-term storage compatibility.

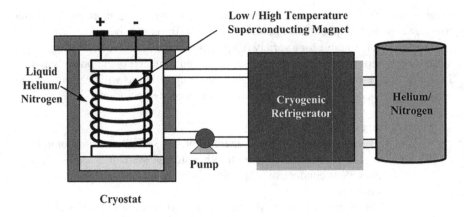

FIGURE 15.8   Superconducting magnetic energy storage.

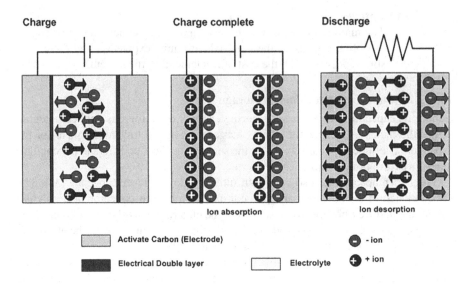

FIGURE 15.9   Supercapacitor energy storage.

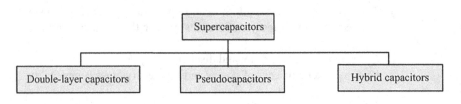

FIGURE 15.10   Types of supercapacitors.

- **Applications**
  1. The supercapacitor is widely used in automobiles for actuators and self-starting engines.
  2. It is better in the field of fluctuating loads such as handheld devices, portable devices, and photovoltaic systems and for stabilizing the power supply.
  3. It is used to reduce power oscillations on the grid.
  4. It is used in electric vehicle battery management systems for supply stabilization and regenerative braking systems [31].

The following development process in the field of supercapacitors is to make the recharging process last a few fractions of seconds, or half the time, making mass energy storage comparable to nickel–metal hydride. The usage of graphene mixed with acetylene black-based supercapacitor commonly known as Super P responsible for higher energy density at room temperature becomes more conductive. The aforementioned discussions have outlined tasks for the future growth of the supercapacitor [32].

15.2.3 THERMAL ENERGY STORAGE (TES)

Thermal energy storage refers to the storage of energy in the form of cold or heat in a storage medium. It includes a storage system as well as separate equipment for the processes of heat injection into the system and heat extraction from the system. This storage medium can be naturally generated, similar to the ground or potholes, or it is feasible to produce it artificially by placing a container over the water tanks to prevent any heat loss or gain [32]. Figure 15.11 depicts various types of TES.

15.2.3.1 Sensible Heat Energy Storage

Thermal energy is stored in sensible heat energy storage systems based on the specific temperature of the substance being used. As a result, the temperature of the material can vary in response to the amount of energy applied, and it does not go through any phase transformations [32]. The quantity of heat that may be held in a

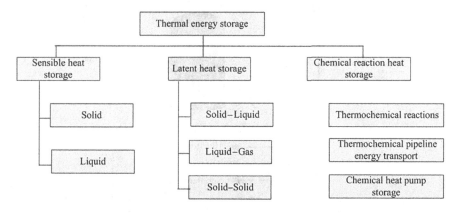

FIGURE 15.11   Types of thermal energy storage.

medium is determined by the temperature variations, the amount of material stored, and the specific heat capacity of the material. Depending on the application, the storage medium for sensible heat storage might be either a liquid or a solid [33].

The liquid medium can be water and oil; the solid medium can be sand, bricks, soil, or rock beds. The input and output of a device are attached along with the container to afford thermal energy for a particular application such as dwellings. This type of thermal storage can be used in houses and offices to provide hot water. Especially in solar heating, water can be used as a storage medium in which heat is stored, called liquid-based systems, as shown in Figure 15.12.

For air-based systems, rock beds can be used as storage systems [34]. Water is used as a storage medium due to its low cost and availability, and it also has excellent heat transfer capabilities. Solid materials can be utilized as a storage medium in TES for low and high temperatures [35]. To minimize the cost, appropriate materials can be used for short-term and long-term energy storage [36]. Figure 15.13 depicts the criteria for selecting the material for sensible heat energy storage.

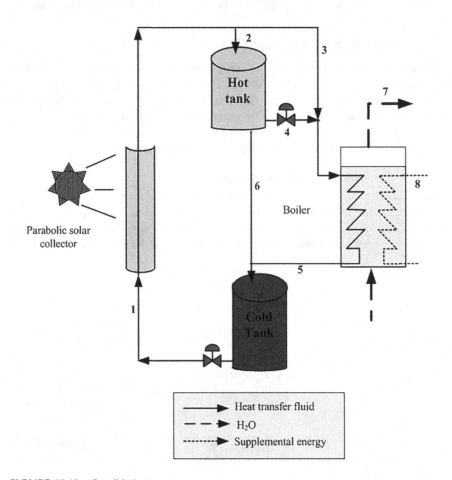

FIGURE 15.12 Sensible heat energy storage

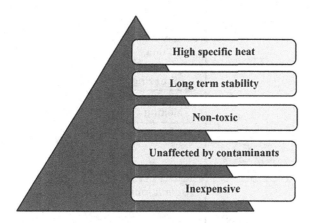

FIGURE 15.13   Sensible heat energy storage—criteria.

- **Advantages of using water as a storage medium**
    1. Wide availability.
    2. Inexpensive.
    3. Non-toxic.
    4. Non-combustible.
    5. High specific heat capacity.
    6. High density.

- **Disadvantages of using water as a storage medium**
    1. Water is either frozen or boiled.
    2. It has limited melting and boiling points.
    3. Water has a corrosive nature.

In order to avoid the above-mentioned problems, chemical additives can be added. Wherever water cannot be used, as an alternative, different liquid mediums, such as molten salts and oils, can be used [36].

15.2.3.2   Latent Heat Energy Storage (LHES)

The process of storing and retrieving thermal energy in a latent heat storage system is based on the latent heat of fusion, in which the storage medium undergoes a phase shift, as shown in Figure 15.14. In response to an increase in temperature at their source, phase change materials (PCMs) experience a breakdown of chemical bonds, resulting in their transformation from one state to another.

The amount of heat released and absorbed during phase transformation from one material to another material is discussed in [37]. To store latent heat, the process of phase transformation has been adapted. The material's phase transformation can be solid–liquid, liquid–gas, or solid–solid. Figure 15.15 depicts the criteria for selecting the material for LHES system, and Figure 15.16 illustrates the classification of LHES materials.

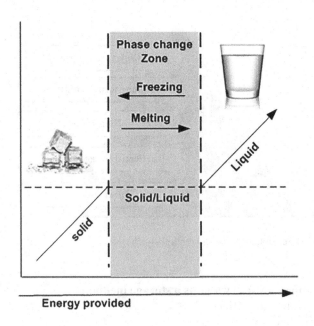

FIGURE 15.14   Latent heat energy storage.

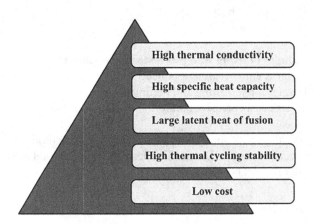

FIGURE 15.15   Latent heat energy storage—criteria.

15.2.3.1.1   Organic Material

1. An organic PCM contains carbon atoms.
2. There are two types of organic material: non-paraffin and paraffin.
3. Chemically synthesized organic PCMs are accessible for a wide range of temperatures, with stability reaching 3,000°C.
4. Examples include wax, hydroquinone, salicylic acid, alpha glucose, and acetamide.
5. Nanomaterials used in paraffin are carbon nanofibers, CuO, and Fe_3O_4.

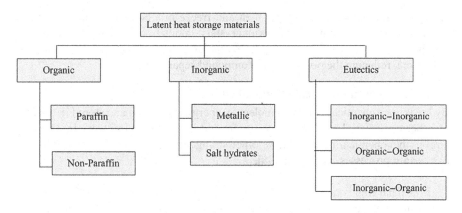

FIGURE 15.16 Classification of latent heat storage.

- **Advantages**
    1. They are chemically stable.
    2. Fusion heat is really high.
    3. There is no trend for supercooling.

- **Disadvantages**
    1. Low thermal conductivity
    2. Mildly corrosive
    3. High cost.

15.2.3.1.2 Inorganic Materials

- Minerals such as salt hydrates, nitrates, and metallics are used to make inorganic PCMs.
- Additionally, inorganic PCMs can be employed at elevated temperatures up to 1,500°C.
- Examples include $NaNO_3$, $NaCl$, $MgBr_2$, and KNO_3.

- **Advantages**
    1. They are of low cost.
    2. The melting point is really high.
    3. Thermal conductivity is high.
    4. Fusion heat is really high.

- **Disadvantages**
    1. Material degradation
    2. Low specific heat
    3. Low cycling stability.

15.2.3.1.3 Eutectics

Eutectic PCMs are made up of two or more compounds that make up a certain percentage of the total composition. Compounds can be organic–organic, inorganic–inorganic, or organic–inorganic. These PCMs melt and freeze in a consistent manner without prompting. They freeze into a close-knit crystal combination, reducing the chances of the chemicals separating. Similarly, during melting, different compounds melt at the same time, reducing the possibility of compound separation.

15.2.3.3 Thermochemical Energy Storage

By breaking a chemical bond in a molecule, energy is released and stored in the form of a thermochemical energy storage system and it is retained in its original form by a reversible chemical reaction. Like other storage systems, thermochemical storage has certain processes, as follows [38].

1. Discharging equation: $C + D \rightarrow A + B + thermal\ energy.$
2. Charging equation: $A + B + thermal\ energy \rightarrow C + D.$
3. Reactants have the ability to store thermal energy for a long time.

From Figure 15.17, it is seen that thermochemical energy storage includes both endothermic and exothermic reactions. It releases water from hydrates and zeolites. There is no heat loss when two thermochemical components A and B are held separately at a particular temperature. For long-term energy storage, thermochemical energy is the most suitable storage method [32].

There are several factors to be considered when choosing the appropriate material for thermochemical energy storage as follows:

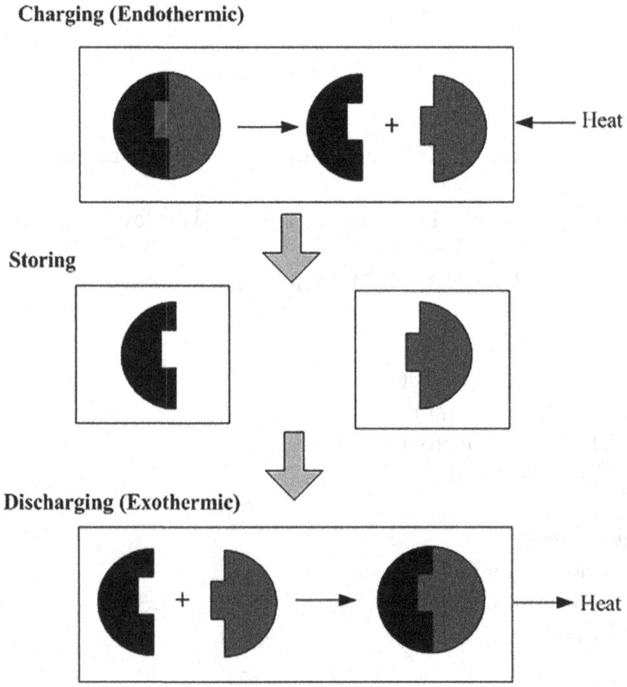

Charging (Endothermic)

Storing

Discharging (Exothermic)

FIGURE 15.17   Thermochemical energy storage.

- Cost
- Designed to withstand more charging, storing, and discharging cycles
- Availability of materials
- Better heat transfer characteristics
- Harmless and non-combustible.

15.2.4 ELECTROCHEMICAL AND BATTERY ENERGY STORAGE

Rechargeable batteries are used as a storage medium in electrochemical energy storage systems [39]. Those rechargeable batteries are flow batteries and secondary batteries. In electrochemical storage systems, chemical energy is produced from electrical energy, and in the reverse process, chemical energy is again altered to electrical energy with better efficiency and fewer physical changes in batteries. These conversion processes in batteries result in a reduction in energy and a decrease in battery life [40]. These batteries serve a dual purpose of storing and releasing electrical energy through a charging and discharging process that emits no harmful emissions and requires no maintenance [41].

15.2.4.1 Flow Batteries

In flow batteries, energy can be conserved by storing it in the form of species that exhibit electrical activity, called electroactive species. This species can be dissolved in tanks that contain a liquid electrolyte, and then it is circulated via electrochemical battery cells, which transform chemical energy into electrical energy. Examples of flow batteries are (i) redox flow batteries and (ii) hybrid flow batteries [32].

15.2.4.1.1 Redox Flow Batteries

The battery's total energy defines the full size of the redox flow battery tank [32]. Redox flow batteries have a stable life cycle, improved efficiency, and power flexibility, and they meet the demand for capacity, making them useful in large grid systems as well as stand-alone applications [42].

Figure 15.18 shows one of the redox flow batteries, called the vanadium redox flow batteries [42]. As shown in Figure 15.18, the vanadium battery consists of two electrolyte solutions in which metal ions are immersed. Those metal ions are pumped into the reverse side of vanadium cells. A membrane separates two electrodes through which electrons are oppositely passed between those electrodes.

The transfer of electrons is done by allowing the protons to pass through the membrane. During the charging process, the flow of current takes place over the electrode, and in the discharging process, active masses in a dissolved state are supplied to the electrodes from the battery tank. A few examples of redox flow batteries are iron–titanium, ferro-chromium, and polystyrene–butadiene systems [32].

15.2.4.1.2 Hybrid Flow Batteries (HFB)

Hybrid flow batteries contain active masses which are stored in both the cell and the liquid electrolyte present in the tank. A HFB is formed by the secondary battery and the redox flow battery. A hybrid flow battery, similar to a redox flow battery, has a set storage capacity that is determined by the entire size of the cell. During charging,

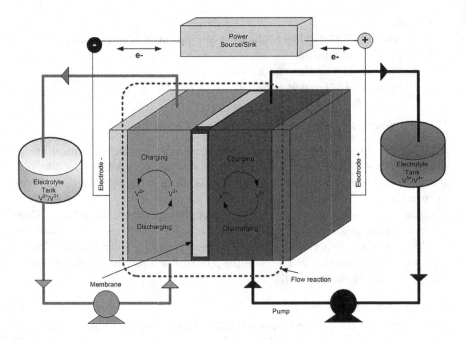

FIGURE 15.18 Vanadium redox flow batteries.

zinc is accumulated over the electrode, and while discharging, the zinc-ion is fed back to the solution [32].

- **Advantages of flow batteries**
  - Long life (15–20 years)
  - Discharge range (4–10 hours)
  - Better efficiency about 60%–70% [42].

In recent days, redox flow and hybrid flow batteries have been used in energy storage of large-scale power applications and in a system that has a storage capacity ranging from a few megawatt-hours to hundreds of MWh for improving the quality of the power, peak leveling of electricity, and enhancing the security of the power supply and non-conventional energy integration [43].

15.2.4.2 Secondary Batteries

Secondary batteries have become the most popular in recent years, owing to their widespread use in portable devices for power supply. They are also used in different applications, such as electronic devices. Secondary batteries store energy in the form of electrical energy, which is converted into chemical energy. Thus, they produce electricity through electrochemical reaction [44]. Secondary batteries contain two electrodes: a cathode and an anode, with an electrolyte solution. Electrodes are separated by a separator, and all are covered by a case-like structure [32]. Secondary batteries have certain characteristics such as good energy density, high specific energy, less resistance, constant discharge profile, a certain range of temperature performance, and better memory effect [45]. But most batteries may be toxic in nature. So, it is very important to consider the environmental impacts while disposing of the battery [32]. Types of secondary batteries are the following:

a. Lead–acid battery
b. Nickel-based battery
c. Zinc-halogen-based battery
d. Metal–air-based battery
e. Lithium battery [14,29,42].

15.2.4.2.1 Lead–Acid Battery (LAB)

LABs have been the most widely used batteries since the 1860s, particularly for energy resources [42]. These batteries are used to start internal combustion engines in vehicles, but they are also used for other purposes, such as backup power in emergencies, storing energy generated by renewable sources, and a storage medium on electric grids. Mainly, these batteries are chosen due to their temperature resistance properties, safety, and their cost [44]. In this battery,

- Positive electrode—PbO_2 (lead dioxide)
- Negative electrode—Pb (lead)
- Electrolyte—H_2SO_4 (sulfuric acid).

Electrochemical reaction for this battery is given by

$$Pb + 2PbO_2 + 2H_2SO_4 \rightleftharpoons 2PbSO_4 + 2H_2O$$

Figure 15.19 depicts the charging and discharging of a lead–acid battery. While discharging, it produces lead sulfate ($PbSO_4$), and in the charging process, it releases

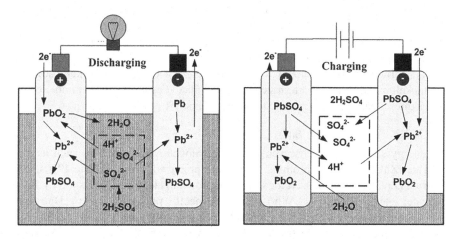

FIGURE 15.19 Charging and discharging of lead–acid battery.

water. The life span of a lead–acid battery is around 6–15 years with 2,000 life cycles. The overall efficiency of the battery ranges from 70% to 90% [44]. Some of the low ratings of LAB are SLI batteries and batteries utilized in UPS with voltage ratings of 6, 8, and 12 V [32].

15.2.4.2.2 Nickel-Based Battery
In nickel-based batteries,

- Negative electrode—iron (Fe)/cadmium (Cd)/zinc (Zn), metal hydride (MH), or hydrogen (H_2)
- Positive electrode—nickel hydroxide ($Ni(OH)_2$)
- Electrolyte solution—potassium hydroxide (KOH).

Based on the negative electrodes, nickel-based batteries are classified into

- Nickel–cadmium (Ni–Cd)
- Nickel–zinc (Ni–Zn)
- Nickel–iron (Ni–Fe)
- Nickel–hydrogen (Ni–H_2)
- Nickel–metal hydride (Ni-MH).

The electrochemical reaction for this battery is given by

$$X + 2NiO(OH) + 2H_2O \rightleftharpoons 2Ni(OH)_2 + X(OH)_2 \quad \text{where, } X = Fe/Cd/Zn$$

$$(M)H + 2NiO(OH) \rightleftharpoons M + Ni(OH)_2$$

$$H_2 + NiO(OH) \rightleftharpoons Ni(OH)_2$$

Figure 15.20 depicts the charging and discharging of a nickel-based battery. Nickel hydroxide $Ni(OH)_2$ is produced during the discharging process, and iron (Fe)/cadmium (Cd)/zinc hydroxide ($Zn(OH)_2$) is generated during the charging process. Costs are high due to the lower specific power, and proper maintenance is required. Nickel–iron and nickel–zinc batteries are not preferable in electric vehicle applications [32].

Their efficiency is about 75%. For powering electric vehicles, nickel–cadmium and nickel–metal hydride batteries can be used since they have a high number of life cycles (more than 2,000) with a better energy density.

15.2.4.2.3 Zinc-Halogen-Based Battery
For energy storage in electric vehicles, zinc-halogen-based batteries such as zinc chloride ($ZnCl_2$) and zinc bromide ($ZrBr_2$) are preferred. From the late 1970s, zinc chloride ($ZnCl_2$) has been used for storage purposes in electric vehicles [8,46]. It has a greater energy density (90 Wh/L) and less specific power (60 W/kg). Due to its greater specific energy, charging capability, and low cost, zinc bromide ($ZrBr_2$) batteries are widely used in electric vehicles for energy storage [47,48]. But nowadays, this battery type of usage is reduced due to a reduction in specific energy (90 W/kg),

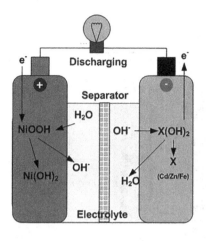 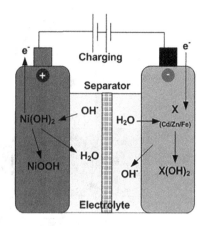

FIGURE 15.20 Charging and discharging of nickel-based battery.

high bromine concentration, and the greater size needed for circulating the electrolyte solution [48,49]. The chemical reaction for the zinc bromide ($ZrBr_2$) battery is as follows:

$$ZnBr_2(aq) \rightleftharpoons Zn + Br(aq)$$

When a chemical reaction takes place in the zinc bromide battery, energy is stored and gets released from the electrodes.

In this battery,

- Negative electrode—zinc (Zn)
- Positive electrode—bromine (Br)
- Electrolyte solution—zinc bromine aqueous solution.

As shown in Figure 15.21, the electrolytic solution is circulated by pumping the electrodes [8,49]. A chemical reaction occurs during charging that deposits zinc on the negative electrode and bromine on the positive electrode. In the discharging process, zinc and bromide ions are released from the electrodes.

15.2.4.2.4 Metal–Air-Based Battery
In this battery,

- Negative electrode—metals such as Li, Ca, Mg, Fe, Al, and Zn
- Positive electrode—air
- Electrolyte solution—alkaline solution.

The most commonly used metal–air-based battery is the lithium–air battery, which is used in electric vehicles. Its specific energy is about 11.14 kWh/kg, without air. When compared to other battery types, the lithium–air battery has a greater energy capacity

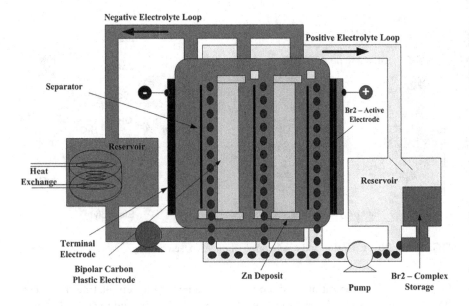

FIGURE 15.21 Zinc–bromide battery.

(more than 100 times) [50]. But one of the drawbacks of this battery is that it may be exposed to fire when the air is combined with humidity [44]. Other battery types, such as lithium–water and lithium–oxygen, are considered to be secondary batteries for ESS [51].

Some of the metal–air-based batteries are given below:

- Calcium–air (Ca–air) battery
- Aluminum–air (Al–air) battery
- Zinc–air (Zn–air) battery.

Due to low cost and greater specific energy, metal–air-based batteries are suitable for reusable energy storage applications. Because of their low material cost and high specific energy, metal–air batteries are suitable for rechargeable storage applications [52,53]. In a metal–air battery, the electrochemical reaction is given by:

$$4M + nO_2 + 2nH_2O \rightleftharpoons 4M(OH)n$$

where
M stands for a metal which can be Ca, Li, Mg, Al, Fe, and Z.
n represents the value obtained when there is a change in the valance shell of metal.

Figure 15.22 shows the chemical reaction of the zinc–air battery charging and discharging processes. While the discharging process, electrons get released so that the electrode in which the zinc is deposited gets oxidized and another electrode in which air is deposited generating hydroxide ions. In the charging process, zinc is

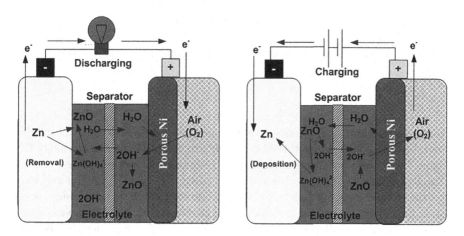

FIGURE 15.22   Charging and discharging of zinc–air battery.

accumulated in the negative electrode and air is discharged into the negative electrode [54].

15.2.4.2.5   Lithium Battery

Among all the secondary batteries, the lithium battery is one of the most prominent ones that have been used for energy storage in recent days. It has certain advantages, such as greater energy density and specific energy with less weight. It has no memory effect and is harmless when compared to a lead battery [55]. One disadvantage of this type of battery is that it is expensive, and in order to be safe, additional protection is required. It also keeps the voltage and current levels constant to improve battery performance, resulting in cell balancing [56]. In high-temperature and medium-temperature applications, these lithium batteries play a major role. Lithium batteries used in high-temperature applications are as follows:

- Lithium–aluminum–iron monosulfide (Li-Al-FeS) and
- Lithium–aluminum–iron disulfide (Li-Al-FeS$_2$).

Among these, lithium–sulfur batteries have better energy capacity and much less weight when compared to other lithium batteries. But it has a low life cycle, needs battery thermal management, and loses energy in order to maintain an operating temperature. In this lithium–sulfur battery [57],

- Negative electrode—Li–Al alloy
- Positive electrode—iron sulfide
- Electrolyte solution—molten lithium chloride potassium.

The electrochemical reaction of lithium–sulfur battery is given by

$$2Li - AlFeS \rightleftharpoons 2Al + Fe + Li_2S$$

$$2Li - Al + FeS_2 \rightleftharpoons 2Al + Fe + Li_2FeS_2$$

Lithium batteries used in atmosphere temperature applications are as follows:

- Li-poly batteries
- Li-ion batteries.

Among these, lithium-ion batteries are most commonly used in both electrical and electronics energy storage applications due to their size, lower weight, and capability [57]. Figure 15.23 shows the chemical reaction of the lithium-ion battery charging and discharging processes. When electrons are emitted during charging, lithium-ions move from the positive electrode to the negative electrode and combine to form lithium atoms. At the time of the discharging process, the process is reversed. This lithium-ion battery was developed, especially for next-generation electric vehicle applications [58,59].

15.2.5 CHEMICAL ENERGY STORAGE

The energy released and stored via chemical reactions is called chemical ESS. Chemical reactions are done by composing chemical compounds in a system which results in the formation of different compounds [10]. Fuel cell technology is one of the chemical energy storage systems. A fuel cell is a technology that converts chemical energy in the form of fuel into electrical energy in a continuous manner [8,49,60]. The source of the energy supplied by the fuel cell and the battery differs. Electricity

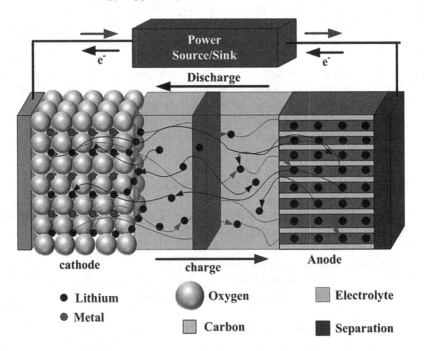

FIGURE 15.23 Chemistry of lithium battery.

is generated in fuel cells by supplying the fuel and oxidant externally, whereas batteries are used internally [49]. A fuel cell can generate electrical energy continuously as long as it receives an active energy source as an input, which is a major advantage. The efficiency of the fuel cell ranges from 40% to 85% [8,60].

Fuel cells are a promising energy generation alternative that reduces the use of fossil fuels and thus CO_2 emissions [60–62]. Liquid and gas are accumulated on the anode of a fuel cell, which serves as fuel. Oxygen and chlorine, which act as oxidants, are discharged at the cathode. The most common fuel cell is a hydrogen fuel cell, which is widely available. Hydrogen and oxygen combine to generate electricity in hydrogen fuel cells. Based on the fuel supplied, hydrogen fuel cells are classified as direct systems or indirect systems. In direct systems, fuel reacts directly with oxidants. In an indirect system, the fuel is converted into hydrogen gas first and the hydrogen gas is fed into the cell for chemical reaction. Based on certain parameters, such as different combinations of fuel and oxidant, electrolytic solution, and temperature, fuel cells are classified into many types as follows [8,49]:

1. Alkaline fuel cell
2. Solid oxide fuel cell
3. Phosphoric acid fuel cell
4. Regenerative fuel cell
5. Molten carbonate fuel cell
6. Direct methanol fuel cell.

The chemical reaction of the hydrogen fuel cell is given by

$$2H_2 + O_2 \rightleftharpoons 2H_2O + \text{Electricity}$$

Figure 15.24 shows the chemical reaction of a hydrogen fuel cell battery [8,60]. In this, hydrogen and electrons are formed when injected hydrogen fuel is detached from the fuel electrode. Once hydrogen fuel is injected, H_2 ions flow toward the electrode, in which oxygen is accumulated through an electrolytic solution.

At the same time, electrons flow through a circuit connected externally to deliver power to the load. Hydrogen ions combine with oxygen and electrons in nearby electrodes to form water. In a reversible process, the electrolyzer is used to separate water molecules into hydrogen and oxygen again. Then the converted hydrogen is fed back to the cell, which produces electricity and water as a by-product. The cycle repeats until there is a need for electricity [49].

15.3   APPLICATIONS OF ENERGY STORAGE SYSTEMS

As per the utilization process concerns, these storage energy resources are deployed in various forms. The power system is one of the areas where the energy storage system plays an important role in the various control mechanisms, such as maintaining high system reliability by controlling generation and transmission based on demand and ensuring the quality of power transmission with the help of ESS, in order to protect the load-side equipment [63]. Figure 15.25 confers different applications of ESS under the concept of power systems.

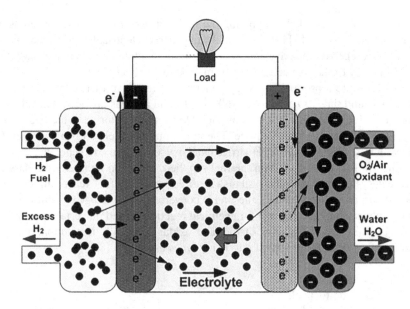

FIGURE 15.24 Hydrogen fuel cell.

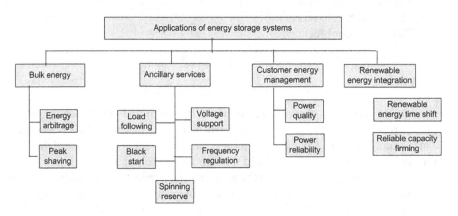

FIGURE 15.25 Classification of ESS applications.

15.3.1 BULK ENERGY APPLICATIONS

To enhance and control the larger variation of power in the grid control system, bulk energy applications are used. The various types of bulk energy applications are discussed below.

15.3.1.1 Energy Arbitrage

Energy trading is aimed at the process of generating profit economically. Producing energy in accordance with the scale of demand may result in a shortfall of some unconditional requirements. As is well known, the cost of generation is quite high, and thus, the idea developed is continuous generation of energy and storing that energy in excess of the requirements. During peak hours, the stored energy is used by selling it at a higher price, allowing energy trading to take place. For example, the microgrid is integrated with renewable energy systems meant for both storing energy and the transmission of power [64].

15.3.1.2 Conservation for Peak Demand

The mechanism of power conservation is the same as discussed in energy arbitrage, but this application does not include it in the process of making it economically beneficial [65]. They are designed to store energy in order to meet peak demands and thus improve overall system capability by compensating power for normal and peak operations using ESS. Storage systems are typically installed on the consumer side in this application, whereas storage systems are kept on the source side in energy arbitrage [66,67].

15.3.2 ANCILLARY SERVICE APPLICATIONS

The utility service's primary goal is to provide assistance to power system components in order to improve the quality of transmission from the source side to the load side. The various applications are described in subsequent points [63].

15.3.2.1 Load Tracking

When looking at the load-side control, load-side variations are unpredictable. They change dynamically in nature according to their operations [68]. The energy storage system, on the other hand, responds quickly to loads. With the inference gained, by synchronizing the energy storage system as an intermediary system for providing power to the load with respect to the variation of load, it acts as a well-balanced system for both the generation-side and the load-side control [66,69].

15.3.2.2 Spinning Reserve

The spinning reserve is backup power that is not utilized during regular functioning [56]. When there is an outage, this storage system provides a path for inoculating power to the system. In this regard, the backup system provides power to the system for one hour. The generation of energy continues to charge up the backup system till it reaches the desired value of power [67,70].

15.3.2.3 Voltage Support

The quality of the transmission of power in the power system between networks is achieved when there is proper stability control of voltage. Variation in transferring potential across networks may result in power loss. To keep the system stable, the management of reactive power plays an important role [71]. The reactive power can be regulated to the required limit by the energy storage system in the nearby fault area [71,72]. With the help of ESS, the possibility of maintaining the voltage is made simpler.

15.3.2.4 Blackout

A blackout occurs as a result of a power outage or the occurrence of certain unanticipated events. This happens in order to deal with the stability of the system [73–75].

After a blackout, there should be a recovery process for maintaining power management and voltage control, called a black start. In this situation, ESS play the role of supplying the required active power for energizing the power lines or initial power for gearing up the power plant for the generation of power [73].

15.3.2.5 Frequency Regulation

Any variation in supply frequency may lead to the misbehaving of the load. The various loads are more sensitive to their frequency, even to slight variations. The energy storage system in the frequency regulator compensates for deviations within the respective ranges. Frequency regulation is classified into three types based on the range of variation and processes, namely

1. Primary reserve control: Compensates between the generation-side and demand-side controls and corrects the frequency within a few seconds of the period for the generator control [76–79].
2. Secondary reserve control: It is a backup for the primary reserve control, avoids the imbalance between the networks, and is ready for frequency rectification within certain limits [77,78].
3. Tertiary reserve control: This should be operated manually; the objective is the same as the secondary reserve control for balancing between networks.

15.3.3 CUSTOMER ENERGY MANAGEMENT APPLICATIONS

Energy management applications determine the standard and reliability of the power delivered to the user.

15.3.3.1 Power Quality

As a result of generation concerns, the energy generated is not the same in every situation. There might be some variation in generation due to environmental issues related to renewable energy systems [80–82] and, commercially, regarding the requirement of raw materials for generations. These variations may lead to defects in the standard quality of power, and generate harmonics and voltage fluctuations [83]. To overcome the problem, the method of energy storage brings an alternative solution for maintaining the standard of energy supplied to the consumer and thereby protecting the equipment [67].

15.3.3.2 Power Reliability

Power reliability is the same as the power quality objective [84,85]. Reliability ensures the efficiency of storage, which involves a longer duration of storage without any loss of energy. In general, the energy storage system in this application should be capable of storing more energy while also providing a consistent supply of power to the user.

15.3.4 RENEWABLE ENERGY INTEGRATION APPLICATIONS

Renewable energy integration into the microgrid makes the entire power system more flexible. The integration system is implemented near the source of the generating stations. ESS are increasingly being utilized in the process of renewable energy storage to compensate for power fluctuations caused by a variety of environmental factors. These applications are utilized in various categories, such as timely usage and capacity firming. Timely usage ensures the utility of the stored energy when there is an outage and stores the energy when the demand is lower. Capacity firming enables the standard output from the storage system for utility purposes to enhance the power quality across the system. Wind, hydro, and solar energy are examples of renewable energy sources, with an emphasis on storage facilities such as FWS, CAES, and PHES [63].

15.3.5 Applications Based on Usage

ESS can be in various locations as per their usage, such as renewable energy integration, power plants, power system networks, and large industrial areas. The utility areas deploy various operations as per the application concerned. The grid-tied system increases storage efficiency, makes the storage device more responsive to changes in load, and allows it to hold large charge and discharge cycles [63,64].

15.3.6 Self-Generation and Utilities

Apart from the usage of grid-tied and grid-interactive systems, the usage of stand-alone storage systems is capable of providing energy at a higher efficiency. This stand-alone storage device receives energy from renewable sources which are incorporated into their buildings [86]. The use of appropriate storage technology is determined based on storage capacity, charge and discharge cycle, and self-discharge rate, and this will reduce the peak load demand [37,87,88].

15.3.7 Transportation

The vehicles used for the transportation of stored energy generated by water and air use gasoline tanks. Electric vehicles have been equipped with ESS such as supercapacitors, flywheels, fuel cells, and batteries (lithium-ion batteries) in recent years [89]. These devices should have a faster discharge power, high energy density, and high efficiency and have the feature of a regenerative braking system. The combined effect of the battery and hybrid energy storage technology, such as supercapacitors and flywheels, improves the vehicle's performance characteristics; they are engine starting, supportive power for actuators in vehicles, and distributive power systems [90]. Vehicles with storage devices capable of self-generation are utilized for proper discharge and load leveling and are used in the case of emergency power failure [91].

15.3.8 General Applications

ESS have a wide range of applications. From the inferred advantages of the current storage systems, this system is the most predominantly used one. The storage system is used in a variety of applications, including solar air heating, cooling, solar, and off-peak electricity storage [41,43]. Flywheels are used on the International

Space Station for earth observation missions. Asphalt concrete containing lime-stone, quartzite, and copper slag is used for thermal storage. Through the process of photon-driven electrolysis of water, we can convert the sun's heat into chemical fuels by producing hydrogen and oxygen, respectively.

15.4 TECHNICAL COMPARISON OF VARIOUS TYPES OF ENERGY STORAGE SYSTEMS

Table 15.1 shows the different energy storage technologies such as MES, TES, CES, ECES, and EES. MES technologies are used in electric trains, electric motors, and other applications that require high efficiency and long life. TES technologies have a long lifetime and high efficiency, and when compared to other technologies, they produce a different output, which is used to combine heat and power. CES technologies, despite their low efficiency, can store energy for a long time period when compared to other technologies. But due to the short storage time, it can only be used for seasonal storage.

In EES technologies, the storage of electric energy is in the form of a charge and can retain power whenever required [92]. This is because supercapacitors and superconductors are used to store energy in the form of electric and magnetic fields. Magnetics have high power and medium energy density storage devices that are used in spacecraft. ECES technologies will be suitable for specific applications based on their power range.

15.5 CHALLENGES AND ISSUES IN DEPLOYING ENERGY STORAGE SYSTEMS

ESS play a major role in power systems and also in the integration of renewable energy with the grid, in transmission systems, as well as in distribution networks. Anyhow, there are certain challenges in energy storage technology, especially in large-scale power applications [32].

Barriers that do not allow energy storage technology to expand worldwide are as follows:

• Price
• Less awareness on energy storage
• Regulations and services that reduce the benefits of energy storage.

Electricity tariffs set by the government will not include any additional credits for calculating the amount of energy saved in energy storage facilities. There is no clear system to reclaim the amount of electricity obtained from energy storage based on the current tariff structure. As a result, energy storage has become less valuable. To change the electricity tariff, approval is needed from the concerned authorities. So, the benefits obtained from energy storage will not be based on the capital cost of investment. It is a major reason that inventors are not attracted to energy storage technology [96]. Another reason is that there will be no proper methodology

TABLE 15.1
Comparison of Energy Storage Devices

Storage Techniques	Output	Period	Power	Efficiency	Maturity
Mechanical energy storage (MES) [93]	Electricity	Very long life time	1–3,000 MW (PHS) Over 80 MW (FES) 35–300 MW (CAES) [63]	76%–85% (PHS) 85%–90% (FES) 70% (CAES) [94]	Mature/demo/ early commercialized
Electrical energy storage (EES)	Electricity	Very long	10–75 kW (SCES) Very high power in MW (SMES)	95% (SCES) 95%–98% (SMES) [94]	Demo/early/ developing
Thermal energy storage (TES) [95]	Thermal	Long life time	0.001–10 MW (SHS) 0.001–1 MW (LHS) 0.001–1 MW (TCES)	30%–50% (overall) [94]	Commercialized/ developing Demo/ commercialized based on material and temperature
Electrochemical energy storage (ECES)	Electricity	Long life time	300–600 (LAB) 0.03–7 MW (RFB) 1,200–4,000 kW (Li-ion) 175–4,000 kW (Li-ion) 500–1,500 kW (Ni–Cd) 300–600 kW (Ni–Cd) 700–2,500 kW (ZnBr) 175–2,500 kW (ZnBr) 100–250 kW (metal–air)	70%–80% (overall) [94]	Mature/demo/ early commercialized
Chemical energy storage (CES)	Electricity	Very long storage time	Varies	40%–50% (overall) [94]	Demo/ developing

for measuring savings and there are no incentives, leading to the belief that energy storage is not a viable alternative to investing in new power plants and transmission lines. This can be remedied by developing new strategies that make good use of energy storage technology and encourage people to invest in energy storage technologies [97].

Furthermore, because of its uniqueness, the higher cost of raw materials used, and the lack of bulk production, the energy storage technology has become the most expensive to implement. Energy storage technologies are less appealing than traditional technologies due to their higher cost, despite the fact that they provide a carbon-free environment. There are a few energy storage types such as PHES that are less expensive, but the main disadvantage of this type is that it offers extra expenses. It can be solved by the government providing funding support, tax incentives, and easing regulations to attract developers [97]. Since the 1970s, energy storage has been developed and the results will be seen in future green technology. Further, there is no knowledge of expanding energy storage for large-scale applications due to the scarcity of decisive data about the cost and capability of energy-saving information. In order for the policymakers to realize the importance of energy storage technology, proper data should be provided by performing a brief study on it.

15.6   FUTURE RESEARCH DIRECTIONS

Even though ESS have a lot of advantages, they still need further developments, which should be cost-effective and have good performance with better solutions. This emerging storage technology needs some in-depth and broad research so that this technology can be accepted worldwide in all aspects (both commercial and industrial areas).

With different energy storage types, continuous research is going on in the field of electrochemical energy storage, which focuses on proposing new raw materials for storage and finding out the appropriateness of various energy storage requirements [60,61]. Batteries are the cheapest energy storage alternative for different applications, with a smaller number of cycles. In particular, lithium batteries play a major role in easily transported electrochemical energy storage technology. But the cost and performance of lithium batteries are limiting factors in their growth in different storage applications. Lithium–air batteries appear to be a promising feature for achieving better performance and are economically viable, even though technical challenges are addressed. Sodium-ion batteries need some more analysis in their field, although they provide low-cost benefits in order to achieve the technical ability for lithium-ion technology [55,56].

The cost of restoration contributes a significant portion of the battery life cycle cost, and it is assumed that the number of cycles of a particular battery can affect the income of a utility, which is an important factor in battery usage. Supercapacitors can be compared with batteries for different energy storage applications and make use of battery features by integrating both and can help in addressing the future of electrical energy storage disputes [1].

In the generation of renewable energy resources, ESS are a much needed method to avoid the fluctuation of power in all fields of renewable energy generation. There should be a proper device or method to be propagated in order to store large amounts of power generation. Several methods are being investigated in the field of research to store more power, including batteries, flywheel storage, magnetic storage, CAES, and PHES [1].

In the consideration of various storage techniques, flow batteries are known to be one of the storage mechanisms for larger amounts of energy. They are quite neglected due to the frequent changes in the electrolyte in the battery, and the electrolyte is very expensive. Other storage methods, such as flywheel storage, are designed for shorter storage periods and higher discharge power. Researchers are drawn to this method because it focuses on reducing power losses both mechanically and electrically, increasing power conversion efficiency, and focusing on materials to strengthen the flywheel rotor.

Among the various storage methods, PHES and CAES are more reliable and capable of large scalable storage, but there is a reduction in round-trip efficiency (power conversion efficiency). To improve production performance in the future, the design of the rotating turbine should be well equipped. On the other hand, in the advancement of CAES, isothermal compressed-air energy storage is a recent technology to maintain a constant temperature for the storage of energy under any variation and is a more efficient method than conventional CAES [32,35,64].

Thermochemical heat storage is the emerging technology to store higher storage density and minimal heat losses and is capable of long-term storage. Much more heat storage analysis is needed, as is the development of storage mediums with improved thermal properties for storing heat with minimal thermal loss. The analysis is based on materials science to neglect adsorbent instability and optimization of temperature during the discharging period. Among these technologies, the hydrogen energy storage system is designed to adapt to changes regularly and provide increased storage capacity in a variety of applications with complete feasibility [38,61].

The effectiveness of various storage mediums was evaluated in order to make them suitable for future circumstances. The subsequent research in the energy storage field is the combination of two or more storage methods in the corresponding applications (synergy), which has the features of higher energy density, faster discharge rate, maintaining an optimal temperature in the medium, etc., thereby delivering standard power to the consumer and maintaining proper energy management [98–101]. When combining storage methods, proper consideration should be given to the power conversion efficiency. With appropriate control and maintenance, it makes the system more reliable and more effective when these storage devices are connected to the power system networks. With the various analyses discussed above, the integration of these storage devices with the grid-connected network makes the entire network smarter and more flexible [102].

15.7 CONCLUSIONS

Practically, there are various types of storage methods involved. The determined storage methods, which are described in this chapter, pave the way for developing different types of storage with unique characteristics. The various challenges encountered during the operation were discussed, and the appropriate problem-solving techniques were expressed in order to restore the normal operation of the system. The research work in the storage field discusses the advancements in techniques to improve the performance of the storage system and increase the power conversion efficiency of the system. Consideration of the various types of storage methods helps in the usage of storage systems for specific applications. The majority of storage systems are used in power system networking applications, such as energy management applications, bulk energy applications, energy trading, peak savings, and load flow studies. These applications help to know about the responsive operation of the storage system. In the future, expanding the storage system with well-equipped and advanced operation as per the application not only improves the performance characteristics, but also reduces the cost and dependency on the availability of fossil fuels and paves the way for the creation of a smarter storage technology with an interactive system known as smart grid.

REFERENCES

1. Koohi-Fayegh, S., & Rosen, M. A. (2020). A review of energy storage types, applications and recent developments. *Journal of Energy Storage*, 27, 101047.
2. Dunn, B., Kamath, H., & Tarascon, J. M. (2011). Electrical energy storage for the grid: A battery of choices. *Science*, 334(6058), 928–935.
3. Whittingham, M. S. (2008). Materials challenges facing electrical energy storage. *Mrs Bulletin*, 33(4), 411–419.
4. Aneke, M., & Wang, M. (2016). Energy storage technologies and real-life applications– A state of the art review. *Applied Energy*, 179, 350–377.
5. Chen, H., Cong, T. N., Yang, W., Tan, C., Li, Y., & Ding, Y. (2009). Progress in electrical energy storage system: A critical review. *Progress in Natural Science*, 19(3), 291–312.
6. Shyam, B., & Kanakasabapathy, P. (2018). Large scale electrical energy storage systems in India-current status and future prospects. *Journal of Energy Storage*, 18, 112–120.
7. Christen, T., & Carlen, M. W. (2000). Theory of Ragone plots. *Journal of Power Sources*, 91(2), 210–216.
8. Chau, K. T., Wong, Y. S., & Chan, C. C. (1999). An overview of energy sources for electric vehicles. *Energy Conversion and Management*, 40(10), 1021–1039.
9. IEC International Electrotechnical Commission. (2011). Electrical Energy Storage. IEC White Paper.
10. Hannan, M. A., Hoque, M. M., Mohamed, A., & Ayob, A. (2017). Review of energy storage systems for electric vehicle applications: Issues and challenges. *Renewable and Sustainable Energy Reviews*, 69, 771–789.
11. Fujihara, T., Imano, H., & Oshima, K. (1998). Development of pump turbine for seawater pumped-storage power plant. *Hitachi Review*, 47(5), 199–202.
12. Rehman, S., Al-Hadhrami, L. M., & Alam, M. M. (2015). Pumped hydro energy storage system: A technological review. *Renewable and Sustainable Energy Reviews*, 44, 586–598.
13. Menéndez, J., Ordóñez, A., Álvarez, R., & Loredo, J. (2019). Energy from closed mines: Underground energy storage and geothermal applications. *Renewable and Sustainable Energy Reviews*, 108, 498–512.
14. Beevers, D., Branchini, L., Orlandini, V., De Pascale, A., & Perez-Blanco, H. (2015). Pumped hydro storage plants with improved operational flexibility using constant speed Francis runners. *Applied Energy*, 137, 629–637.
15. Vigna, K. R., Gomathi, V., Ekanayake, J. B., & Tiong, S. K. (2019). Modelling and simulation of variable speed pico hydel energy storage system for microgrid applications. *Journal of Energy Storage*, 24, 100808.
16. Yang, C. J., & Jackson, R. B. (2011). Opportunities and barriers to pumped-hydro energy storage in the United States. *Renewable and Sustainable Energy Reviews*, 15(1), 839–844.
17. Vasel-Be-Hagh, A., Carriveau, R., & Ting, D. S. K. (2013). Energy storage using weights hydraulically lifted above ground. *International Journal of Environmental Studies*, 70(5), 792–799.
18. Liu, H., & Jiang, J. (2007). Flywheel energy storage—An upswing technology for energy sustainability. *Energy and Buildings*, 39(5), 599–604.
19. Lemofouet, S., & Rufer, A. (2006). Hybrid energy storage system based on compressed air and super-capacitors with maximum efficiency point tracking (MEPT). *IEEJ Transactions on Industry Applications*, 126(7), 911–920.

20. Budt, M., Wolf, D., Span, R., & Yan, J. (2016). A review on compressed air energy storage: Basic principles, past milestones and recent developments. *Applied Energy*, 170, 250–268.
21. Krawczyk, P., Szabłowski, Ł., Karellas, S., Kakaras, E., & Badyda, K. (2018). Comparative thermodynamic analysis of compressed air and liquid air energy storage systems. *Energy*, 142, 46–54.
22. Abdo, R. F., Pedro, H. T., Koury, R. N., Machado, L., Coimbra, C. F., & Porto, M. P. (2015). Performance evaluation of various cryogenic energy storage systems. *Energy*, 90, 1024–1032.
23. Antonelli, M., Barsali, S., Desideri, U., Giglioli, R., Paganucci, F., & Pasini, G. (2017). Liquid air energy storage: Potential and challenges of hybrid power plants. *Applied Energy*, 194, 522–529.
24. She, X., Peng, X., Zhang, T., Cong, L., & Ding, Y. (2019). Preliminary study of Liquid Air Energy Storage integrated with LNG cold recovery. *Energy Procedia*, 158, 4903–4908.
25. Faraji, F., Majazi, A., & Al-Haddad, K. (2017). A comprehensive review of flywheel energy storage system technology. *Renewable and Sustainable Energy Reviews*, 67, 477–490.
26. Bolund, B., Bernhoff, H., & Leijon, M. (2007). Flywheel energy and power storage systems. *Renewable and Sustainable Energy Reviews*, 11(2), 235–258.
27. Baker, J. (2008). New technology and possible advances in energy storage. *Energy Policy*, 36(12), 4368–4373.
28. Toodeji, H. (2019). A developed flywheel energy storage with built-in rotating supercapacitors. *Turkish Journal of Electrical Engineering and Computer Science*, 27(1), 213–229.
29. Ali, M. H., Wu, B., & Dougal, R. A. (2010). An overview of SMES applications in power and energy systems. *IEEE Transactions on Sustainable Energy*, 1(1), 38–47.
30. ASHRAE. (2007). *Thermal Storage. ASHRAE Handbook: HVAC Applications.* American Society of Heating, Refrigerating and Air Conditioning Engineers (ASHRAE), Inc., Atlanta, GA.
31. Naish, C., McCubbin, I., Edberg, O., & Harfoot, M. (2007). *Outlook of Energy Storage Technologies.* Policy Department Economic and Scientific Policy, Brussel.
32. Mahlia, T. M. I., Saktisahdan, T. J., Jannifar, A., Hasan, M. H., & Matseelar, H. S. C. (2014). A review of available methods and development on energy storage; technology update. *Renewable and Sustainable Energy Reviews*, 33, 532–545.
33. Domański, R., Jaworski, M., & Rebow, M. (1995). Thermal energy storage problems. *Journal of Power Technologies*, 79, 1–25.
34. Farid, M. M., Khudhair, A. M., Razack, S. A. K., & Al-Hallaj, S. (2004). A review on phase change energy storage: Materials and applications. *Energy Conversion and Management*, 45(9–10), 1597–1615.
35. KHINMYAMYA. Encapsulation of phase change materials for heat storage. 2003.
36. Fernandez, A. I., Martínez, M., Segarra, M., Martorell, I., & Cabeza, L. F. (2010). Selection of materials with potential in sensible thermal energy storage. *Solar Energy Materials and Solar Cells*, 94(10), 1723–1729.
37. Sharma, A., Tyagi, V. V., Chen, C. R., & Buddhi, D. (2009). Review on thermal energy storage with phase change materials and applications. *Renewable and Sustainable Energy Reviews*, 13(2), 318–345.
38. Abedin, A. H., & Rosen, M. A. (2012). Closed and open thermochemical energy storage: Energy-and exergy-based comparisons. *Energy*, 41(1), 83–92.
39. Hiroshima, N., Hatta, H., Koyama, M., Yoshimura, J., Nagura, Y., Goto, K., & Kogo, Y. (2016). Spin test of three-dimensional composite rotor for flywheel energy storage system. *Composite Structures*, 136, 626–634.

40. Linden, D., & Reddy, T. B. (2002). *Linden's Handbook of Batteries* (Vol. 4). McGraw-Hill, New York.
41. Ibrahim, H., Ilinca, A., & Perron, J. (2008). Energy storage systems—Characteristics and comparisons. *Renewable and Sustainable Energy Reviews*, 12(5), 1221–1250.
42. Noack, J., Roznyatovskaya, N., Herr, T., & Fischer, P. (2015). The chemistry of redox-flow batteries. *Angewandte Chemie International Edition*, 54(34), 9776–9809.
43. Wei, X., Xu, W., Huang, J., Zhang, L., Walter, E., Lawrence, C., Vijayakumar, M., Henderson, W.A., Liu, T., Cosimbescu, L., & Li, B.. (2015). Radical compatibility with nonaqueous electrolytes and its impact on an all-organic redox flow battery. *Angewandte Chemie International Edition*, 54(30), 8684–8687.
44. Akinyele, D. O., & Rayudu, R. K. (2014). Review of energy storage technologies for sustainable power networks. *Sustainable Energy Technologies and Assessments*, 8, 74–91.
45. Xing, Y., Ma, E. W., Tsui, K. L., & Pecht, M. (2011). Battery management systems in electric and hybrid vehicles. *Energies*, 4(11), 1840–1857.
46. Cui, S. M., Lu, Y., Song, J. P., Wang, J. F., & Ding, W. F. (2013). Study on Zn-PANi battery characteristics used for electric vehicles. In *Advanced Materials Research* (Vol. 724, pp. 1374–1378). Trans Tech Publications Ltd, Switzerland.
47. Lai, Q., Zhang, H., Li, X., Zhang, L., & Cheng, Y. (2013). A novel single flow zinc–bromine battery with improved energy density. *Journal of Power Sources*, 235, 1–4.
48. Manla, E., Nasiri, A., Rentel, C. H., & Hughes, M. (2009). Modeling of zinc bromide energy storage for vehicular applications. *IEEE Transactions on Industrial Electronics*, Elesiver, 57(2), 624–632.
49. Yao, L., Yang, B., Cui, H., Zhuang, J., Ye, J., & Xue, J. (2016). Challenges and progresses of energy storage technology and its application in power systems. *Journal of Modern Power Systems and Clean Energy*, 4(4), 519–528.
50. Lee, J. S., Tai Kim, S., Cao, R., Choi, N. S., Liu, M., Lee, K. T., & Cho, J. (2011). Metal–air batteries with high energy density: Li–air versus Zn–air. *Advanced Energy Materials*, 1(1), 34–50.
51. Lu, Y. C., Gallant, B. M., Kwabi, D. G., Harding, J. R., Mitchell, R. R., Whittingham, M. S., & Shao-Horn, Y. (2013). Lithium–oxygen batteries: Bridging mechanistic understanding and battery performance. *Energy & Environmental Science*, 6(3), 750–768.
52. Spanos, C., Turney, D. E., & Fthenakis, V. (2015). Life-cycle analysis of flow-assisted nickel zinc-, manganese dioxide-, and valve-regulated lead-acid batteries designed for demand-charge reduction. *Renewable and Sustainable Energy Reviews*, 43, 478–494.
53. Atwater, T. B., & Dobley, A. (2011). Metal/air batteries. *Linden's Handbook of Batteries*. McGraw-Hill Education, New York.
54. Akhil, A. A., Huff, G., Currier, A. B., Kaun, B. C., Rastler, D. M., Chen, S. B., Cotter, A. L., Bradshaw, D. T. & Gauntlett, W. D. (2013). *DOE/EPRI 2013 Electricity Storage Handbook in Collaboration with NRECA* (Vol. 1, p. 340). Sandia National Laboratories, Albuquerque, NM.
55. Zhang, Y., Wang, L., Guo, Z., Xu, Y., Wang, Y., & Peng, H. (2016). High-performance lithium–air battery with a coaxial-fiber architecture. *Angewandte Chemie International Edition*, 55(14), 4487–4491.
56. Gallagher, K. G., Goebel, S., Greszler, T., Mathias, M., Oelerich, W., Eroglu, D., & Srinivasan, V. (2014). Quantifying the promise of lithium–air batteries for electric vehicles. *Energy & Environmental Science*, Elsevier, 7(5), 1555–1563.
57. Lim, H. D., Lee, B., Zheng, Y., Hong, J., Kim, J., Gwon, H., Ko, Y., Lee, M., Cho, K., & Kang, K. (2016). Rational design of redox mediators for advanced Li–O_2 batteries. *Nature Energy*, 1(6), 1–9.
58. Yu, S. H., Lee, D. J., Park, M., Kwon, S. G., Lee, H. S., Jin, A., Lee, K. S., Lee, J. E., Oh, M. H., Kang, K., & Sung, Y. E. (2015). Hybrid cellular nanosheets for high-performance lithium-ion battery anodes. *Journal of the American Chemical Society*, 137(37), 11954–11961.

59. Nazri, G. A., & Pistoia, G. (Eds.). (2008). *Lithium Batteries: Science and Technology*. Springer Science & Business Media, Cham.
60. Rashid, M. H. (Ed.). (2017). *Power Electronics Handbook*. Butterworth-Heinemann, United Kingdom.
61. Capasso, C., & Veneri, O. (2015). Laboratory bench to test ZEBRA battery plus supercapacitor based propulsion systems for urban electric transportation. *Energy Procedia*, 75, 1956–1961.
62. Hosseinifar, M., & Petric, A. (2016). Effect of high charge rate on cycle life of ZEBRA (Na/NiCl$_2$) cells. *Journal of the Electrochemical Society*, 163(7), A1226.
63. Palizban, O., & Kauhaniemi, K. (2016). Energy storage systems in modern grids— Matrix of technologies and applications. *Journal of Energy Storage*, 6, 248–259.
64. Bragard, M., Soltau, N., Thomas, S., & De Doncker, R. W. (2010). The balance of renewable sources and user demands in grids: Power electronics for modular battery energy storage systems. *IEEE Transactions on Power Electronics*, 25(12), 3049–3056.
65. Makansi, J., & Abboud, J. E. F. F. (2002). Energy storage: The missing link in the electricity value chain. Energy Storage Council White Paper.
66. Eyer, J., & Corey, G. (2010). Energy storage for the electricity grid: Benefits and market potential assessment guide. *Sandia National Laboratories*, 20(10), 5.
67. Wu, X., Wang, X., Li, J., Guo, J., Zhang, K., & Chen, J. (2013, May). A joint operation model and solution for hybrid wind energy storage systems. In Zhongguo Dianji Gongcheng Xuebao/*Proceedings of the Chinese Society of Electrical Engineering* (Vol. 33, No. 13, pp. 10–17). Chinese Society of Electrical Engineering, China.
68. Mohd, A., Ortjohann, E., Schmelter, A., Hamsic, N., & Morton, D. (2008, June). Challenges in integrating distributed energy storage systems into future smart grid. In *2008 IEEE International Symposium on Industrial Electronics* (pp. 1627–1632). IEEE, Cambridge, UK.
69. Hirst, E., & Kirby, B. (1999). Separating and measuring the regulation and load-following ancillary services. *Utilities Policy*, 8(2), 75–81.
70. Brown, P. D., Lopes, J. P., & Matos, M. A. (2008). Optimization of pumped storage capacity in an isolated power system with large renewable penetration. *IEEE Transactions on Power Systems*, 23(2), 523–531.
71. Eyer, J. M., & Corey, G. P. (2010). Energy storage for the electricity grid: Benefits and market potential assessment guide: A study for the DOE Energy Storage Systems Program (No. SAND2010-0815). Sandia National Laboratories.
72. Katiraei, F., Iravani, M. R., & Lehn, P. W. (2005). Micro-grid autonomous operation during and subsequent to islanding process. *IEEE Transactions on Power Delivery*, 20(1), 248–257.
73. Feltes, J. W., & Grande-Moran, C. (2008, July). Black start studies for system restoration. In *2008 IEEE Power and Energy Society General Meeting-Conversion and Delivery of Electrical Energy in the 21st Century* (pp. 1–8). IEEE, Pittsburgh, Pennsylvania, Allegheny.
74. Laaksonen, H., & Kauhaniemi, K. (2008). Control principles for blackstart and island operation of microgrid. In *Nordic Workshop on Power and Industrial Electronics (NORPIE/2008)*, June 9–11, 2008, Helsinki University of Technology, Espoo, Finland.
75. Sun, W., Liu, C. C., & Liu, S. (2011, July). Black start capability assessment in power system restoration. In *2011 IEEE Power and Energy Society General Meeting* (pp. 1–7). IEEE, Detroit, MI, USA.
76. Bradbury, K. (2010). *Energy Storage Technology Review* (pp. 1–34). Duke University.
77. Energinet, D. K. (2012). Ancillary services to be delivered in Denmark Tender conditions. *Technical report*, Energinet, Fredericia.
78. Vuorinen, A. (2008). *Planning of Optimal Power Systems*, 1st ed. Ekoenergo Oy, Espoo.
79. Galiana, F. D., Bouffard, F., Arroyo, J. M., & Restrepo, J. F. (2005). Scheduling and pricing of coupled energy and primary, secondary, and tertiary reserves. *Proceedings of the IEEE*, 93(11), Denmark, 1970–1983.

80. Boyle, G. (2004). *Renewable Energy*, 3rd ed. Oxford University, Oxford.
81. Bevrani, H., Ghosh, A., & Ledwich, G. (2010). Renewable energy sources and frequency regulation: Survey and new perspectives. *IET Renewable Power Generation*, 4(5), 438–457.
82. Atwa, Y. M., El-Saadany, E. F., Salama, M. M. A., & Seethapathy, R. (2009). Optimal renewable resources mix for distribution system energy loss minimization. *IEEE Transactions on Power Systems*, 25(1), 360–370.
83. Mundackal, J., Varghese, A. C., Sreekala, P., & Reshmi, V. (2013, June). Grid power quality improvement and battery energy storage in wind energy systems. In *2013 Annual International Conference on Emerging Research Areas and 2013 International Conference on Microelectronics, Communications and Renewable Energy* (pp. 1–6). IEEE, Kanjirapally, India.
84. Brown, R. E. (2017). *Electric Power Distribution Reliability*. CRC Press, Boca Raton, FL.
85. Moreno-Muñoz, A. (Ed.). (2007). *Power Quality: Mitigation Technologies in a Distributed Environment*. Springer Science & Business Media, Berlin/Heidelberg, Germany.
86. Del Pero, C., Aste, N., Paksoy, H., Haghighat, F., Grillo, S., & Leonforte, F. (2018). Energy storage key performance indicators for building application. *Sustainable Cities and Society*, 40, 54–65.
87. Tatsidjodoung, P., Le Pierrès, N., & Luo, L. (2013). A review of potential materials for thermal energy storage in building applications. *Renewable and Sustainable Energy Reviews*, 18, 327–349.
88. Arteconi, A., Hewitt, N. J., & Polonara, F. (2012). State of the art of thermal storage for demand-side management. *Applied Energy*, 93, 371–389.
89. Thackeray, M. M., Wolverton, C., & Isaacs, E. D. (2012). Electrical energy storage for transportation—approaching the limits of, and going beyond, lithium-ion batteries. *Energy & Environmental Science*, 5(7), 7854–7863.
90. Cao, J., & Emadi, A. (2011). A new battery/ultracapacitor hybrid energy storage system for electric, hybrid, and plug-in hybrid electric vehicles. *IEEE Transactions on Power Electronics*, 27(1), 122–132.
91. Khaligh, A., & Li, Z. (2010). Battery, ultracapacitor, fuel cell, and hybrid energy storage systems for electric, hybrid electric, fuel cell, and plug-in hybrid electric vehicles: State of the art. *IEEE Transactions on Vehicular Technology*, 59(6), 2806–2814.
92. Energy Storage: Program Planning Document, U.S. Dept. Energy, Washington, DC, USA, 2011.
93. Gagnon, Y., & Landrya, M. (2015). Energy storage: Technology applications and policy options. In *2015 International Conference on Alternative Energy in Developing Countries and Emerging Economies* (Vol. 79, pp. 315–320), Abu Dhabi, the United Arab Emirates.
94. Nadeem, F., Hussain, S. S., Tiwari, P. K., Goswami, A. K., & Ustun, T. S. (2018). Comparative review of energy storage systems, their roles, and impacts on future power systems. *IEEE Access*, 7, 4555–4585.
95. Kyriakopoulos, G. L., & Arabatzis, G. (2016). Electrical energy storage systems in electricity generation: Energy policies, innovative technologies, and regulatory regimes. *Renewable and Sustainable Energy Reviews*, 56, 1044–1067.
96. Kaplan, S. M. (2009). Electric power storage. Congressional Research Service.
97. Elkind, E. N., Weissman, S., & Hecht, S. (2010). *The Power of Energy Storage: How to Increase Deployment in California to Reduce Greenhouse Gas Emissions*. Center for Law and the Environment, Berkeley, and Environmental Law Center, UCLA, July.

98. Zhang, L. L., Wang, Z. L., Xu, D., Zhang, X. B., & Wang, L. M. (2013). The development and challenges of rechargeable non-aqueous lithium–air batteries. *International Journal of Smart and Nano Materials*, 4(1), 27–46.

99. Sharma, P., & Bhatti, T. S. (2010). A review on electrochemical double-layer capacitors. *Energy Conversion and Management*, 51(12), 2901–2912.

100. Qi, D., Liu, Y., Liu, Z., Zhang, L., & Chen, X. (2017). Design of architectures and materials in in-plane micro-supercapacitors: Current status and future challenges. *Advanced Materials*, 29(5), 1602802.

101. Choi, C., Kim, S., Kim, R., Choi, Y., Kim, S., Jung, H. Y., Yang, J. H., & Kim, H. T. (2017). A review of vanadium electrolytes for vanadium redox flow batteries. *Renewable and Sustainable Energy Reviews*, 69, 263–274.

102. Abdel-Wahab, M., & Ali, D. (2013). A conceptual framework for the evaluation of fuel-cell energy systems in the UK built environment. *International Journal of Green Energy*, 10(2), 137–150.

16 Cyber and Theft Attacks on Smart Electric Metering Systems
An Overview of Defenses

Asif Iqbal Kawoosa and Deepak Prashar
Lovely Professional University

CONTENTS

16.1　INTRODUCTION

A smart grid is an electrical infrastructure much like a legacy power grid with scalable and pervasive two-way communications, timely control capabilities, large-scale integration of distributed resources, and efficient use of resources. The smart grid provides the features of pervasive smart monitoring technologies, automatic equipment fault sensing, and self-healing. It uses sensors, controlling devices, and monitoring and recording devices over a robust, efficient, and secure ICT infrastructure. Unlike traditional electric grids, it provides a two-way communication compared to traditional grids where it flows primarily in one direction only, i.e., from service provider to consumer. The smart grid has the ability to use IoT in remote controlling and monitoring of advanced metering infrastructure (AMI).The features such as "wireless automatic meter reading" (WAMR), monitoring of power system stability, distributed energy resources optimization, and applications of demand response system make it purely an intelligent infrastructure. It has licensed-exempt bands, security technologies, and interoperability in different smart grid communication standards. Given some of these salient features, smart grids without any doubt are the future power infrastructure of the energy world. Smart grids have changed the definition of customer from consumer to prosumer (producer plus consumer). The customer can also produce some amount of the electricity at home by using solar panels, etc., and can sell it to service providers or directly to other customers using a peer-to-peer transaction approach. The customer can also purchase the electricity from the service provider or from any other customer when needed. These trials have already taken place in many countries such as the United States, the UK, Germany, and India. The use of smart meters has given service providers a better and efficient way for recording meters and issuing bills as per consumption.

As a smart grid has the ability to connect millions of consumers and devices in a network, it demands robustness, reliability, and security. The secure system helps the smart grid to proliferate and build trust among the customers. Security is thus undoubtedly one of the major challenges in the smart grid systems because of the long-range communication on open networks.

16.2　VULNERABILITIES OF ATTACKS IN SMART ELECTRICAL INFRASTRUCTURES

Electrical infrastructures are also known as critical infrastructures. The critical infrastructures are very vital infrastructures of a country. Any sort of disruption in it may lead to a debilitating effect on the economy, development, and public service system of the country, as electricity is used in almost everything in our lives from transport system, agriculture, communication, health to financial services. Its security hence becomes the prime priority for a country. Cybercriminals, hackers, and terrorists always try to attack the national infrastructures due to their malicious intentions and even to get control on the energy monitoring and controlling systems for personal gains. Cybercriminals or terrorists may launch any attack with the aim to damage and disrupt such installation also to get global attention and intervention. An attack on electric grid systems can plunge a whole city, state, or even a country into darkness. To save these critical infrastructures is on a government's priority list regardless of the party affiliations and manifestos. Despite the security upgrades, new threats are emerging and the challenge is increasing to protect these CI from cyberattacks since the last decade. So the adversary for such attacks is predicated by the ability to combat the attack by active intervention of defenders and have a proactive approach in estimating the attack dynamics. Security is also demanded when customers indulge in stealing the electricity by many means either illegally by-passing the metering systems, or manipulating the readings in the metering systems. The authorities also need to be extremely cautious about these possible theft attacks on the energy metering system installed either at electricity distribution poles or in the customers' premises. Service providers and governments are highly concerned about this economic loss and are spending time and money in making the system highly secure, reliable, and robust. Cybersecurity isn't about keeping hackers away and maintaining the data privacy of customers; rather, it is a resilience of whole grid infrastructure. Security in the smart grid means ensuring everything is safe and well managed when things might inevitably go out of control [1].

16.3 SUMMARY OF SAFETY MEASURES NEEDED IN SMART GRIDS TO DETECT AND PREVENT CYBERATTACKS

The traditional electrical grids are changing into smart grids [2]. The conventional electricity grids evolving to smart grid is an integration of conventional electricity grids and information and communication technology (ICT). Due to its heterogeneous infrastructure, it can be challenging to design a state-of-the-art security mechanism for safeguarding the communications within the different layers and across the smart grid infrastructures [3,4]. This integration has compelled service providers (utility) to purchase security solutions (both software and hardware) for upgrading the existing security after such integration. To enhance the performance, security and monitoring, controlling, and handling the customers' demands, the electricity companies are bound to upgrade and improve the security [5].

The security measures needed within and outside the smart grid infrastructure are to have:

- Availability of uninterrupted power supply within consumer demands.
- Integrity of communicated data.
- Confidentiality of consumers' data, emphasis on smart grid's vulnerabilities, and the kind of attacks that can occur and their preventions and solutions.

- Preventive and reactive safety measures for intrusions and mitigation for vital systems in smart grids. Unlike conventional electric infrastructures, the effect of an attack on a smart grid—a vital infrastructure—could be devastating and consequences could be more damaging. Thus, it is needed to proactively detect the potential threat for avoiding the damaging consequences.
- Automating the safety evaluation in smart grids.
- The self-healing property of the smart grid after any system error or minimal attack.
- Measures of safety and resiliency in smart grids

16.4 ARCHITECTURE OF SMART GRID

A smart grid assembles all the collected data coming from multiple metering devices, smart sensors, and other appliances and manages them efficiently for demand response systems, integration of the electrical resources distributed over large areas, and communication within the components. This communication is mostly transmitted through the wireless channels (less secure) for communicating critical data in such infrastructures.

The architecture of smart grids is a highly distributed and hierarchical architecture based on the reference model suggested by the National Institute of Science and Technology (NIST), the USA (Figure 16.1).

The NIST framework discussed in Ref. [6] emphasizes on the intelligent monitoring system, intelligent control systems, demand-side response energy management systems, and others, depending on the security of the system. The cyber layer in the smart grid is more often exposed to many security threats. The other things such as

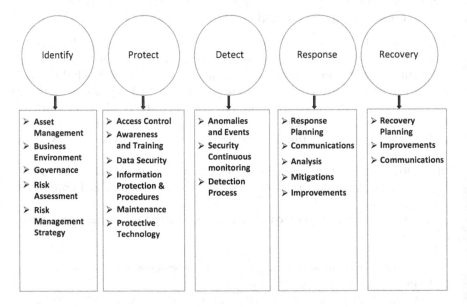

FIGURE 16.1 NIST architecture.

time-sensitive operational requirements and large number of stakeholders add to the security concerns. The advanced metering infrastructure (AMI) also possesses the high risk of cyberattacks and is the easy path for intruding into the system. The work in Ref. [6] discusses the proposed solutions put forth by the researchers in context with cybersecurity. Some other research that has been done over the years in context with cybersecurity has been reviewed in this book chapter.

The smart grid uses two levels of communications: One is HAN (home area network) and the other is WAN (wide area network). The HAN is an emerging technology in the smart grid sector for serving homes with digital application solutions and energy management. It empowers the consumers and allows the smart grid infrastructure to benefit the home owners directly, thus assisting the utility in managing peak electric demand. With HAN providing a centralized access to multiple devices, communication is achieved efficiently and intelligently and energy optimization is achieved by way of managing grid load, energy monitoring, and controlling by utilities and consumers also achieve energy saving by getting energy alerts and pricings dynamically and regularly. A WAN is used for large geographical areas. The model of National Institute of Standards and Technology (NIST) helps smart grids provide resourceful information by studying the "use cases" and identify interoperability interfaces for enhancing the cybersecurity strategies. In the NIST model, logical domains such as bulk generation transmission, distribution, and customer domain use two-way communication for power and information flow, whereas the logical domains such as markets, service providers, and operations use only one-way communication for managing power flow in a smart grid infrastructure.

16.5 PROPOSED SOLUTIONS BY SOME OF THE RESEACHERS ON ATTACKS

Risk-managed security compliance doesn't guarantee the foolproof security against the attacks. But, selection of the appropriate security measures, in the context of the threat and vulnerability, allowing the security platform to mitigate the risk and manage the threat. The objective of such security is to create a risk-managed process strategy that succeeds in mitigating the vulnerabilities of threats and reducing the potential impact to an acceptable level. We need to understand that there is always some percentage of risk involved and it can never be 100% safe. We need to manage the residual risk. A resilient cybersecurity infrastructure resists risk and reduces the overall impact of the attack. A balanced approach is the key; that is, having complexity of security doesn't qualify for the strong security and also having less security increases vulnerabilities and leaves it open and exposed to threats. Selecting the appropriate security measures, in the context of the threat, vulnerability, and consequence, allows the security platform to mitigate the risk and manage the threat. Countermeasures are only implemented when they are necessary to mitigate the threat. Security, like safety, is everyone's responsibility. The security program relies on the efforts of all the components to secure the smart grid.

In the research paper [7], information is provided related to networking architecture of smart grids. The research paper [7] defines the simple architecture of network distribution in a hierarchical network system, which provides the information about the functionality of smart grids because of different architecture in smart grid, information flow is modified as per methods and designs. The hierarchical network system improves the reliability of the network (reliable network delivers secure, real-time messages and non-real-time monitoring and management), communication models, protocol suite, traffic model (periodic traffic flow for constant monitoring), timing requirements, etc.

The guidelines provided by the NIST suggest encryption and authentication for improving the security, integrity, and reliability of the data across the network because if an attacker gets control of the AMI, he has then a path to intrude into the otherwise secure system and can cost highly in terms of energy theft and data manipulation. Therefore, more secure encryption and authentication system is needed to be installed for the prevention of such intrusion. Also, the security system that is used for risk assessment and management identifies the critical assets, finds its vulnerability, and on that assessment reports, prioritizes the risk prevention and the risk mitigation tactics. So the risk management frameworks and secure system architecture in the smart grids are strengthened against the intruder attacks, inadvertent errors, system failures, and natural disasters. One of the main and important- device used in smart grid is intelligent electronic devices (IEDs). This device is as susceptible to attacks as other devices. IEDs are used to remotely control the components of the smart grid but their remote working makes them exposed to a variety of cyberattacks, e.g., spoofed remote device. Various firmware/software are used for resisting the tampering and remote unauthorized access to the IEDs. One such firmware embedded in it is called code-auditing. A study by LeMay et al. presented an architecture named cumulative attestation of kernel for securing the networked embedded system. The authors in Ref. [7] also discussed cyberattack that can enter into the physical network despite having an intrusion detection system in place used as a data security in state estimation method. It says that for enhanced security and reliability, intrusion detection system algorithms must be enabled on each component of the system. The authors of the paper [8] evaluated the security threats in a top-down approach in the communication infrastructure of smart grid. They categorize attacks into three types as denying the availability of the network, compromising the data privacy and data integrity. They concluded the study by saying that the network performance gets degraded if system comes under aDoS attack (Figure 16.2).

The capabilities of the smart grid include the enhanced reliability, accessibility, and efficiency of data over communication networks. Some researchers propose the smart grid security solution in using a cryptographic solution such as public key infrastructure and trusted components.

The smart grid can be called an intelligent grid as it has the capability to communicate in both ways between its units and has improved grid interoperability. The smart grid features that have revolutionized the power system are the automatic transfer of data between metering equipment and the substation, and generation, distribution, and transmission substations with increased accuracy, resilience, improved monitoring and control, and physical security, as discussed in Ref. [9]. The intelligence in the smart grid comes by embedding a microprocessor in most units of the system with a robust operating system. Further, independent components with smart sensors are connected to have access to their own unit through the operating system

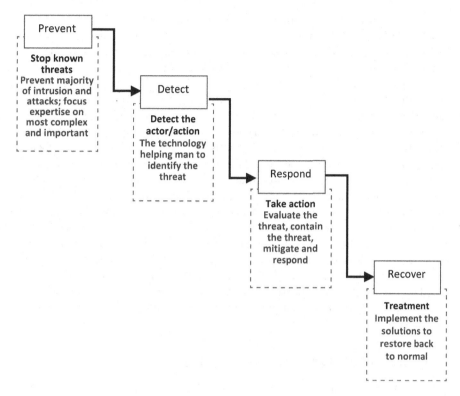

FIGURE 16.2 Intelligence-driven security in smart grid system as defined in NIST architecture.

and a communication path to neighboring units to access operating conditions of circuit breaker, as mentioned in Ref. [10]. The integration of large-scale distributed infrastructure has triggered vulnerability not only to physical attacks and natural disasters, but mostly to cyberattacks and is considered as a major concern today. The attacks lead to huge economic loss, infrastructural failure, energy theft, privacy theft, and uncertainty in operating procedures and can greatly reduce efficiency and reliability of the SGs. Now the resilience of smart grid against cyberattacks is very crucial due to the infrastructural loss it can create on national sensitive infrastructures. It needs to have auto-healing or self-healing ability, control and advanced sensing, and timely detection system as mentioned by the authors in Ref. [11]. The cybersecurity of the smart grid is the most concerning issue due to its job of protecting the precious national infrastructures. It has been noticed that some components are more prone to the attacks of hacking or are more vulnerable compared to the other components such as AMIs and exposed wireless mesh structure networks. The attacks that have grown by adding new technologies in the smart grids are discussed in detail in [2–5].

A considerable number of attacks have taken place and are also expected to occur on smart grids, which are mentioned by the authors in Ref. [12]. The cyberattacks on the smart grid infrastructure are classified in terms of *attacker type* or *attack type*.

The authors of Ref. [13] discussed the four main major components of SG—AMI, SCADA, PHEV, and communication standards and protocols. The data at AMIs are manipulated by masquerading as a legitimate meter management system and hacking the data for monetary or non-monetary gain by mischief. The hacker then manages to kick out the legitimate user by changing control commands. To shield the AMIs from these attackers, the data communications between smart meters and service providers are to be made highly secure by using software and hardware security components, encryption, digital signature, firewall, and access methods and trusted environment platforms need to be implemented. Zigbee, an IEEE 802.15.4 framework used in AMIs, is an easy target for hackers to exploit it by hacking. The most common attack on Zigbee is called Killer Bee attack. Moreover, these components need to be updated appropriately to ensure closing of doors that hackers use to enter the working environment. In addition, smart meters need to be installed at secure positions in consumers' premises to avoid easy alteration. The authors of Ref. [13] discussed the various levels in security issues in the smart meter and measures proposed by the researchers to save smart meters from such cyberattacks. Few security issues in smart meters are also identified in this chapter. The cyberattacks are broadly classified into three categories: attacks on network, attacks on physical hardware, and lastly, attacks on communication data.

16.6 ATTACKER TYPES IN SMART GRIDS

The attack carried out by the attacker to damage different levels can be either a conscious attack or an unconscious attack. Some attacks are carried out to reduce electricity billing, for which they attack their AMI or billing system. Attackers are grouped according to the motivation of attack and target of attack. Some of them try to create warfare; some are related to terrorism, trying to damage the national critical infrastructures to disrupt government functioning. Some are disgruntled employees who try to harm the development and try to exploit the weak points; some are for industrial espionage or economic reasons. Attackers are also grouped in terms of their knowledge about the infrastructure or communication channel likely to be attacked, for example professional attacker and amateur attacker. Some are terrorists that want to create law and order problems. Some are competitors, employees, or customers [14]. Non-malicious attackers are those who only perform attacks, but they not harm the system or does not have any personal or commercial motives. They can be people who want to check their ability to break the security levels, script kiddies, those who are learning to solve some puzzle, hobbyists, etc. Malicious attackers only do attacks to harm.

16.7 ATTACK TYPE

Attacks can be physical, environmental, and cyber. The physical threat is an unauthorized access to physical infrastructure in the system. Environmental threats deal with natural threats such as natural calamity, flood, fire, and extreme heat and cold, and the leading cyberthreats are manipulation, sabotage, and espionage.

The two types of attacks that are considered to compromise the security of smart grids are passive attacks and active attacks. Passive attacks are attacks that violate the confidentiality principle of the security. They attack the system to know the configuration, normal behavior, and architecture and usually are unnoticeable as no data change is found. So these attacks are prevented instead of detected. Examples of such attacks are eavesdropping attack and network traffic analysis.

Active attacks are attacks that are aimed to modify the existing data in the system or inject malicious data into the system to violate availability, confidentiality, and integrity principles of security.

The vulnerability in the system can be in software or in a computational unit, which leads to the failure in security of data. To address the securities, we need to study the attack and its behavior so that preventive measures are taken. One of the major attacks that happened in 2011 was Stuxnet. Stuxnet is considered a zero-day attack. A zero-day attack is an attack that remains unknown till it finishes the job of infecting the system completely, as mentioned in Ref. [15]. Stuxnet attack was an intentional attack, which was carried out on a SCADA (Supervisory Control and Data Acquisition) to harm the nuclear program of Iran. It was carried out on a nuclear power plant (Bushehr Power Plant) in Iran. It was considered the first malware attack directly attacking the control system of the critical infrastructure. The PLCs (programmable logic controllers) were compromised to collect information and cause fast-spinning centrifuges to tear themselves apart. Stuxnet comprised of three modules, a worm, a link-file, and a rootkit. It was carried out on Windows machines, and it infected them with the help of an USB stick [16]. The virus attack was smart enough to find the high-speed centrifuges that are used to enrich fuel (nuclear in this power plant) and took control of the control system, causing physical malfunctioning. Stuxnet was considered the most dangerous advanced malware as it was hybrid consisting of worm, rootkit, virus, and Trojan and was 20 times more dangerous to any normal malware. It stepped in the system as a worm using Trojan; then, it started discovering the target system; next, it was destroying that target; and finally, it was evading to trap or detect. Stuxnet succeeded in attacking the system, but it also led to strengthening the cybersecurity in the smart grids. This also led to the strong need of embedding the full-fledged cybersecurity system to the smart grids. The existing cybersecurity systems still have few loopholes, which concludes that the electrical energy is still vulnerable to cyberattacks with current defense systems.

16.8 CLASSIFICATION OF CYBERATTACKS

Attacks in the smart grids target the targeting CIA triad in smart grids: the availability of energy to the consumers, its integrity, and finally, the confidentiality or privacy of the consumer's identity (Table 16.1).

The cyberattacks directly or indirectly compromise the CIA triad of the system. The various malwares are worms, Trojans, viruses, spywares, trapdoors, backdoors, etc. Generally, we see most types of attacks among them or combination of them. Some types of malwares that include logic bomb, backdoor, and trapdoor are embedded in the system via pre-installed software and are used to initiate attacks. DOS and DDOS attacks are carried out to delay the data transmission and damage the data

TABLE 16.1

Classification of Cyberattacks that Block the CIA Principle in the Smart Grids

S. No.	Confidentiality	Reference
1	Man-in-the-middle attack	[17]
2	Password pilfering	[18]
3	Spoofing	[19]
4	Unauthorized access	[20]
5	Traffic analysis	[21]
6	Eavesdropping	[22]
	Integrity	**Reference**
1	Tampering, wormhole	[23]
2	Replay	[24]
3	Spoofing	[19]
4	Data injection	[14]
5	Time synchronization	[14]
6	Data modification	[25]
	Availability	**Reference**
1	Wormhole, flooding	[23]
2	Puppet attack	[26]
3	DoS/DDoS	[27]
4	Jamming	[28]
5	Buffer overflow	[29]

while transmission between nodes in a smart grid [30]. Man-in-the-middle attack, software exploitation, and message replays are called spoofing attacks. One of the attacks called a jamming attack gets connected with the communication channel only to show its impact. The attack that remains unknown till it infects and completes the infection process is called a zero-day attack. In August 2003, most parts of the USA and Canada experienced power shutdown possibly due to a cyberattack to a system software. It affected 500 lakh people and 62,000 MW load causing serious implications. In September 2003, some parts of Italy and Switzerland also experienced a power blackout in which 5,600 lakh people suffered blackouts and brownouts. This happened due to a failure in the system and due to the error in software, which resulted in communication failure [31].

16.9   THE SET OF DETECTION AND PREVENTION SYSTEMS IN SMART GRID

16.9.1   INTRUSION DETECTION SYSTEM IN SG

The communication network of the smart grid works the same as Internet, but these communication channels are superimposed on the power lines system having the standard of 802.15.4, 802.11, and WiMAX. This network is also open and exposed like Internet and is vulnerable to attacks from external agents.

The authors of Ref. [32] proposed a smart grid distributed intrusion detection system (SGDIDS), which is an embedded intelligent module called the analyzing module (AM). It works in multiple layers in the smart grid. The AMs deployed on each layer of the smart grid are modules trained on relevant data on each level and are able to communicate and improve methods of detection of malicious data. Multiple AMs are deployed at various layers such as home area networks (HANs), wide area networks (WANs), and neighborhood area networks (NANs), where security is improved by using techniques such as support vector machine (SVM) and artificial immune system (AIS) for detecting malicious traffic. This methodology has shown promising results in improving the security by identifying malicious data and also selecting the optimal routes for communication traffic. Multiple simulations for this method have shown the positive effectiveness in improving security.

16.9.2 Use of Frequency Monitoring in Communication

The authors of [33] proposed an evolution path for communication in SGs on cognitive band, i.e., on unlicensed spectrum of radio waves for standard interoperability and cybersecurity. The communications in SG transmission systems use wide-area frequency monitoring networks (FNETs) with local clients and use wireless WAN or Internet technologies with remote data centers. The FNETs use frequency disturbance recorders (FDRs) and sensors for measuring values of the phase angle, the values of amplitude, and also the values of frequency in the system. FNETs are used in the communication that has large real-time data. FNETs also facilitate the wide-area monitoring used to observe key values in the electrical circuit. Frequency distribution recorders are put in use as sensors are used to measure the key values in the system, such as phasor values of amplitude, phase angle, and frequency of the voltage [34].

16.9.3 Cognitive Radio-Based WRANs for Communication in SG

The cognitive radio technologies are used in SG communication due to their scarce unlicensed frequency band. The delay in sensing the critical data and shifting band to cognitive spectrum included in IEEE 802.22 makes it challenging. So a new concept of using dual-radio band architecture is also proposed by researchers in smart grids for receiving and transmitting data [33]. One of the LTE-A and most popular 4G BB technologies called WiMAX is used as communication network at San Diego, Australia, Texas, etc., and is called a hybrid communication network [35].

16.9.4 The Wireless Sensor Networks (WSNs)

These form the very essential part of smart grid establishment. The WSNs have the ability to monitor embedded electric utility systems and help in the implementation of embedded electric utility monitoring systems, remote monitoring systems, etc. [33]. WSNs help in keeping track of energy flow and its accounting and monitoring of all the operational communications. The SG is built on an open protocol network so that it has the option of interoperability where both legacy and new grids have coexistence and preserve its communication security in all circumstances. SG offers two-way communication capability with remote control and monitoring of components and efficient use of electricity. To realize these functionalities, WSNs form the essential part in its implementation. The WSNs help to achieve the remote monitoring of systems, fault sensing and healing, remote wireless automatic meter reading, and the management of distributed energy resources. The WSNs can be used in implementing functionalities of SG due to their rapid development, flexibility, and economically feasible advantages [33].

16.9.5 FALSE DATA INJECTION ATTACKS

During this attack, hackers enter in the system and inject modified data or false data rather than launching the physical attack on the power system. It is basically targeted to modify only the data reported from sensors of electric power rather than modifying real electric power flow. A severe data attack uses system error tolerance to get away from data attack detection as it by-passes the traditional bad data detection method. To prevent such attacks, a framework is particularly customized for cross-validating the cyber- and physical defense system strategies. In physical power systems, state estimation methods are used to decrease the observation errors and also to detect false data injection in the system. One of the open-source network intrusion prevention systems called Snort is deployed in the network system to monitor and detect the abnormal transmission of data traffic in the communication network system as mentioned in Ref. [36].

16.9.6 ABNORMAL TRAFFIC-INDEXED STATE ESTIMATION (ATSE)

ATSE is used to detect the possibility of data attacks in smart grids. These attacks modify the real and actual data sensed by sensors rather than modifying the actual or real power flow. The ATSE quantifies every discrete event in the power system. The bad data injection attack is detected by *cyber–physical fusion methods*.

16.9.7 USE OF IOTS

IoTs are the most preferred and reliable method to be used in data security and communication among the various components in the smart grid because cybersecurity deployments need to use uniform approaches in different segments for maintaining connectivity. The most essential components are customer's privacy, trusted components across the networks, and software security. That is why the solution proposed by the researchers at present times is the use of the IoT (Internet of things)-based conceptual model, which is discussed in Ref. [27].

16.10 IPv6 AND IBM MODEL

The communication data exchanged in smart grids are divided into two parts:(i) informational data, the data used to calculate electricity usage bill, consumer's personal information and details, tagging, logging, etc., and (ii) operational data, i.e., the real-time voltage and current values, transformer's feeder loads, transformer's tap changer positions, relay status, and circuit breaker status. Security for the network, software modules, and the communication that monitors and analyzes the data is a major concern. The key solutions offered are using DoS detection and DoS mitigation methods, data security methods, key management system, and network security protocols. Despite using network security protocols, compatibility in heterogeneous networks still remains a challenge. The problem occurs when a high-computing device has to communicate with a low-computing device and when a high-power device communicates with a low-power device, and vice versa. The authors of [37] discussed the IBM model that shows the data aggregation may cause severe data security vulnerabilities due to the fact that uniform protocol cannot be used in all segments of the network. It is suggested by them that a model similar to IBM with a little change and a single segment is used for communication and each segment is connected to the Internet and communicates separately without aggregating data through multiple links to the application layer. This solves the problem in data aggregation; the data are sent through 4G or Wi-Fi networks. The IPv6 features are capable of solving the integration problem as compared to low-power wireless personal area networks, which is discussed in Ref. [27].

In the article [38], the authors proposed a robust and efficient infrastructure with a level of interdependencies as the interdependency of subcomponents reduces the efficiency and reliability if one of the subcomponents is attacked or hacked during a cyberattack. The rest of the components are not affected and are operational even if others are attacked in the real-time frame.

16.11   USE OF 5G

Using 5G and beyond networks, smart grid becomes a combination of operational technology (OT) with the evolving information technology (IT) for improved efficiency of energy utilization. The authors of Ref. [39] provided a brief overview of working of machine-to-machine (M2M) communication in the smart grid. They also discussed the security issues in M2M communication in the smart grid. This book chapter presents a few solutions to cyberthreats in M2M communications. It discusses the use of the latest and evolving 5G technology in the smart grids and its benefits on its efficiency and also provides a solution to the security concerns in the smart grid. Telemetry, now M2M, was the initial technology, but limited to radio signals for transmitting and measuring data from sensor sources of Internet, mobile devices, WSNs, etc. [40]. The M2M can nowadays be compared with a much evolving technology called Internet of things (IoT). IoT devices are being installed at various places with the capability to communicate on a real-time basis. These can be sensors, smart IEDs, etc., and can communicate and access through the Internet. M2M communication is started when the data from a remote device are either received or sent using a wired/wireless network, power line communications carriers, or communication satellites to other M2M applications where more processing is required [39]. The security concerns such as manipulation and DoS attacks on smart devices such as IEDs with malicious intent can be prevented by using real-time secure communication on the 5G network. The 5G will allow secure communication between the customer and service provider with the capability to automatically identify the energy meters. The use of 5G smart meter applications also provide data encryption for enhanced security [39].

16.12   FAULT DETECTION IS ALSO POSSIBLE WITH PHASOR MEASUREMENT UNITS

PMUs measure the amplitude and phase angle of the sine waves in electric power for dynamic remote area monitoring. Now, the data monitoring is less frequent and stability in terms of latency and reliability is also more in protection components. However, with 5G monitoring abilities and securities and time synchronization, more imperative and metering applications require highly enhanced data encryption techniques for security and user privacy laws. The deployment of the 5G network and its underlying technology makes it suitable for the smart grid.

16.13   USING MACHINE LEARNING AND DEEP LEARNING FOR CYBERSECURITY

Machine learning (ML) techniques have been used in cybersecurity in recent times. Machine learning (ML) and deep learning (DL) have become an important and valuable tool for various applications in the cybersecurity field. A study [41] reviewed DL architectures in cybersecurity applications by highlighting the work and contributions presented by the different researchers in this context. This chapter reviews cyber cyberthreats such as intrusion detection and spam detection. This chapter discusses malware detection methods, phishing detection texts, and botnet detection methods. It also checks for network traffic analysis, CAPTCHA, and binary analysis using DL architectures for cybersecurity, including cryptography, cloud security, biometric security, and IoT security. It also discusses DL architectures, application of NLP, importance of signal and image processing techniques, importance of big data (BD), importance of transfer learning approach, reinforcement learning approach, importance of cryptography, and blockchain technology for detecting and preventing the cyberattacks in the systems. The DL is applied in developing cyberthreat situational awareness using DGA, URL, e-mail, and security log data analysis and CAPTCHA. The CAPTCHA technique identifies whether the user is a human or a bot. A description of ML and DL architectures is provided, and these techniques help to handle very large amount of data and to extract important hidden features more accurately. Also a description of various deep learning software libraries, importance of shared tasks in cybersecurity, and the issues in existing cybersecurity solutions are also given in Ref. [44]. Other than deep learning architectures in cybersecurity, the importance of explainable AI, transfer learning, reinforcement learning, etc., are also discussed.

16.14   GAME-THEORETIC APPROACH USED TO MODEL ATTACKS AND DEFENSES

A solution proposed in [42] suggests the application of the game theory model to model three levels of power system starting with generation and till distribution through transmission. So the levels being modeled are power plants, transmission, and distribution networks. The defenses and attacks in smart grid network security are modeled using the game-theoretic approach. It is considered that both the attacker and the defender at all the three levels give best responses and use the equilibrium strategies. Sensitivity analyses of the equilibrium strategies conclude that when an attacker during an attack reaches a threshold, the efforts by the defender in defending that attack increases. In contrast, the method used for showing this effort level is never affected by the principle of probability of equilibrium. Some highlights of modeling are also thoroughly analyzing the cyberthreats to the smart grid networks. A simple network is taken, and a single attacker or threat is inflicted in trying to do maximum damage while the defender automatically maximizes its efforts to protect the system at no additional costs. The model provides the interactions of the defender and an attacker in accordance with the game theory model. This model has an optimal attack and defense strategy that simulates the power plants, the transmission system, and the distribution network system. One of the research problems here is to consider continuous levels of attack and defense. It does generate a more complex problem for the defender and the attacker. It allows decentralized defense where defender of power plant networks is different and for transmissions it is totally different and are considered as separate decision-makers. Also, a partial failure of networks is also needed to be considered.

16.15 USE OF A PRACTICAL GROUP BLIND SIGNATURE SCHEME

The privacy leakage is a major concern in the smart grids. The research also needs to focus on the protection of privacy, and it has two main aspects: data aggregation (aggregation based on mathematical model and algorithm) and anonymous user authentication. The data aggregation helps to obtain fine-grained user consumption information without revealing consumers identity. Anonymous authentication fails to identify the illegal or malicious user. A scheme called group blind signature scheme is used to accomplish conditional anonymity. It means the anonymity of user is possible to be removed or revoked if the consumption data are tempered by that malicious user. It also achieves the integrity of consumption data by using a verification method called homomorphic encryption (HE), which does transfer very little data for communication overhead between smart meter and control room. The authors of Ref. [43] showed that the proposed scheme is scalable and efficient. And the future research is to accomplish the anonymous reward when a consumer's power consumption meets certain standards by using the blockchain.

16.16 BLOCKCHAIN METHOD

Using the blockchain is a safe system of recording information in such a way that it is almost impossible to edit, cheat, or hack the system. A blockchain technology is essentially a digital ledger of transactions (DLT) that is capable of duplicating and distributing transactions across the network of computer systems in blockchain without relying on a third party. Each block contains a number of transactions in the chain, and on every new transaction on the blockchain, a record is added to every participant's digital ledger with an immutable cryptographic signature called hash. This technology has also been used in smart grids for cybersecurity and is discussed in the research paper [45]. This chapter presents the insights into ideas, techniques, and architectures of implementation of blockchain in the cybersecurity of smart grid. It also proposes a solution for providing cybersecurity to smart grids and the data protection by storing data in the blockchain network that is then validated in the network by the peers for load forecasting, demand response, and local redistribution. To achieve cybersecurity, the blockchain-based applications and services are studied and many applications are used on blockchain DAPPS layer for providing security over cloud and also to establish control of smart grid. Hence, storage, transfer, security, and control all are achieved by using blockchain in a smart grid. However, the study in [45] still shows great challenges that exist in smart grids while using blockchain for cybersecurity [46].

16.17 A SITUATION-AWARE SCHEME FOR EFFICIENT DEVICE AUTHENTICATION

One of the networks that are not directly controlled by the utilities in smart grids and are most vulnerable in nature are home area networks (HANs). HANs are thus controlled by a mechanism called device authentication, as discussed in Ref. [16]. This chapter proposed an efficient scheme called situation-aware scheme for device authentication that utilizes the security risk information assessed by smart home systems having a feature of situational awareness. A suitable authentication protocol based on the assessed security risk level having adequate security level and computational and communication complexity is selected. The design of the proposed scheme is verified through formal verification, and informal security analysis is also proposed in this paper [16]. The performance analysis of the proposed scheme shows efficient results in terms of computational and communication costs.

16.18 KEY MANAGEMENT PROTOCOL FOR SECURE COMMUNICATION

A detailed discussion is given in Ref. [47] about the data exchange which takes place as per the standard of IEC 62351 used in communications, and data are studied to get the assessment of security weaknesses. The sensitive control data can't afford any transfer latency so it needs to be protected for latency. D GOOSE (Generic Object-Oriented Substation Event), which is a communication message for protection control having time criticalityand communication between merging units, and IEDs in a substation over Ethernet known as Sampled Values (SV) protocols are also at risk due to cyberthreats. The study in [17] proposed the use of protocols such as hash and private key, which facilitates the key agreement between substation and the data center at the right time to overcome the security weaknesses. The proposed scheme consists of two phases: registration and key agreement. First, the information exchange takes place between the substation and the data center through a secure channel in registration phase, while the session key is generated in the second phase between the substation and data center. Also, authentication is reviewed in the key agreement stage between the two entities using two protocols, i.e., GOOSE and SV. A tool called AVISPA is used to analyze the security by testing different types of attacks; it investigates the communication privacy and authentication between two entities; also, few manual attacks are carried out to test its safety.

16.19   DETECTION OF ATTACK IN SMART METER USING PROTECTION METHOD

The authors of Ref. [44] discussed cybersecurity issues such as natural disasters(such as volcanoes, tornadoes, floods, and hurricanes), failures (due to Internet or network cable breakdown, errors, and human failures), intentional attacks (those attacks done by hackers and cybercriminals, who attack this infrastructure for different reasons such as energy theft, fraud, sabotage, and vandalism), accidental damages (caused due to unreliable use of information, unintentional exchange of data, or exploring inadequate design), financial risks (an unreliable, obsolete system at the design), and reputation risks (personal information theft) that face one of the biggest problems threatening the smart grid. The protection of smart meters against fire and managing its safety is also mentioned. The algorithm is proposed that exchanges the messages point to point in a distributed model using the standard communication model of networks. Also an intelligent device called guard meter is proposed, which can be used at each microgrid and a controller to raise alarm or communicate in case of fire. A controller management algorithm (CAM) diagnoses and monitors each smart meter's temperature instantly in the smart grid and reports to guard meter in case of high temperature or fire. And the second algorithm smart meter algorithm (SMA) shows the working information of the smart meter at varied temperatures [44].

16.20   DETECTION OF ATTACK USING MACHINE LEARNING METHODS IN SMART METERS DATA

Smart electric meters dynamically measure how much electricity we use and are the newest upgraded energy meters that replace regular electricity meters. All the countries are gradually replacing the traditional meter with smart meters. All the energy companies are making serious efforts for bringing this change in the electrical power system infrastructure. This change is important for providing 24/7 supply of electricity to consumers, measuring and recording consumption reading remotely and automatically, logging information about the use of energy on a real-time and dynamic basis, and getting remote information about the use of electricity by each and every consumer. Smart metering systems have grown over the last decade, and this growth has also given immense challenges regarding its security. ICT integration in these metering systems has opened a few more gates for intruders or hackers to get their hands on the sensitive data of electricity companies and users. This is demanding more security as compared to what is being offered till now in these electrical infrastructures. IT-compliance frameworks demand secure transmission, calculation, storage, and regulation of data. The primary concern should be to keep them secure from damages and detect abnormalities. These are the main mantras that are demanded by the companies from cybersecurity teams. These teams implement frameworks for compliance and implement a set of tools and processes to detect potential threats, and to avoid, defend, and protect the assets of an enterprise.

16.21 DEFICIT DUE TO ENERGY THEFT

The energy theft is a major concern and a severe problem in today's world as huge power gets stolen by the illegal consumers without getting noticed. This also imbalances the demand–supply gap to a great extent. In India, energy theft in electricity is calculated to be $17 billion, which is 30%–40% of total generation of electricity and is a major loss to the utility companies and electricity managing boards. In the USA also, the estimated theft of electricity calculates to a financial loss of $6 billion annually, which is about 1%–3% of the country's total revenue. Similarly, in other countries too, there is an estimation of huge losses to the utility companies due to electricity theft. Hence, anti-theft mechanisms in the systems are being developed and improved every day. On the other hand, the consumers are also finding new methods for stealing the electricity, such as by-passing the meter completely and tampering with the CT coil of the energy meter to record false or less energy than the actual energy consumed. Developing a model that can detect the energy theft in the energy meters effectively and accurately is a big motivation for researchers as energy theft causes a huge financial loss to the developed and developing nations across the globe. Accurate and efficient methods for predicting and detecting energy theft by machine learning techniques have gained more attention across the nations. Many ML-based methods such as support vector machine (SVM) and neural network (NN) are already in use for energy theft detection, but still a single algorithm or technique in use isn't able to do the job with full satisfaction. So instead of using a single learner or technique, it is better to add multiple learners or techniques called ensembles to detect the energy theft. An ensemble is a combination of multiple weak learners to form one strong learner with higher performance.

16.22 ENSEMBLE-BASED METHODS

These are the widely applied techniques for real-time data classification. Ensemble-based learning is considered more effective and has good performance as compared to a single model. Machine learning finds a single model that fits best for finding a solution to the problem rather than making any new model and taking it for granted that it will fit our problem. An ensemble model consists of multiple models and forms a model on the average of these multiple models. Ensemble learning systems (ELS) train a number of machine learning algorithms to make a final decision. This technique makes use of human experience for discovering and integrating many ideas to solve any problem. Among many ensemble algorithms, bagging and boosting are the most common techniques in use at present. As the power system generates large amounts of data with many dimensions in it, deep learning could have been more suitable to be used in data processing and data mining on the data generated by power systems. But some challenges are found in ML-based detection; that is, no explicit features exist in ML-based ETD techniques and, thus, due to this reason, handcrafted features and shallow-architecture classifiers suffer from low accuracy rate in theft detection. Also these DL-based techniques consider consumers' energy consumption as static and are unable to take internal time series nature and the external influence features and factors into consideration. The machine learning-based approaches are found to be the most reliable solutions to ETD, but still, they also suffer from few limitations which need to be addressed. The major limitations of the ETD methods are the instability and inability to model imbalance data, sensitivity toward initial parameters, and poor generalization methods. These problems are overcome by using ensemble techniques. These methods improve classification by aggregating multiple models' results. An individual model's misclassification of data and effect of noisy data is mitigated by the aggregation process. The imbalance between benign and theft data is a big issue. The theft data are as low as about only 10% of the total data in real-world samples. Considering these circumstances, it is proposed in this study that an ensemble classifier-based detector is highly recommended for energy theft detection. Also, there are some major drawbacks in ML-based ETD that the performance degrades with imbalanced data. The imbalance problem is due to less theft data compared to the normal data in the datasets. Ensemble learning methods are categorized into two streams. The first stream uses the technique which takes the results of multiple estimators and finds their combined average result. The combined average result of the estimator is better than that of the individual estimators. Examples are bagging and random forests. The second technique combines the multiple weak estimators and gets a strong estimator. It also reduces the bias. Examples of this stream are AdaBoost, CatBoost, LightGBM, and XGBoost.

16.23   ENSEMBLE ML ALGORITHM

Data preparation is the first and foremost step in anomaly detection. The accuracy of any theft detection system depends on the quality of the dataset. The collected dataset shall be checked for missing values. A k-fold cross-validation will be used to divide the dataset into two or more k-fold validations, and then, the performance of the system shall be checked using the validation techniques.

- **Implementation**: Ensemble learning is a method used where multiple inducers (also known as base learners) are combined to make a decision. It is used typically in supervised ML problems. An inducer takes a set of labeled examples as input and then gives a model (i.e., a classifier or repressor) as output that generalizes these examples that combine multiple inducers to make a decision, typically in supervised machine learning tasks. An inducer, also referred to as a baselearner, is an algorithm that takes a set of labeled examples as input and produces a model (e.g., a classifier or repressor) that generalizes these examples. This output model can then be used for testing new unlabeled examples. An ensemble base learner can be any machine learning algorithm, such as decision tree, neural network, and linear regression. The ensemble learning proves beneficial because by combining multiple models a single inducer is produced that negates the errors and weaknesses of the single inducers and gives an overall better prediction performance than any single model. The reasons ensemble methods improve predictive performance are (i) overfitting avoidance, (ii) computational advantage, and (iii) representation. Ensemble methods mitigate the following: class imbalance, concept drift, and curse of dimensionality.

Given a dataset of n examples and m features: $D = \{(x_i, y_i)\}$ ($|D| = n$, $x \in Rm, y_i \in R$); an ensemble learning model φ uses an aggregation function G that aggregates K inducers $\{f_1, f_2, \ldots, f_k\}$ toward predicting a single output as follows:

$$\widehat{y}_i = \varphi(x_i) = G(f_1, f_2, \ldots f_k)$$

where $\widehat{y}_i \in R$ for regression problems and $\widehat{y}_i \in Z$ for classification problems. The concepts are used to implement the ensemble ML with multiple learners for better results.

Algorithm

Input: I, T, S
(where I is a weak inducer, T is the no. of iterations, and S is a training set).
Output: M_t, α_t; where $t = 1, \ldots, T$
for each t in 1,..., T do
        Build a classifier M_t using inducer I and distribution D_t.
        $\epsilon_t \sum_{i: Mt(xi) \neq yi} D_t(i)$
        $\epsilon_t < 0.5$ *then*
        T t-1;
        exit the loop
        end
else

        α_t ½ log(1- e^t)/e^t
        $D_{t+1}(i) = D_t(i) \cdot e^{-\alpha t.yt.Mt(xi)}$
        normalize D_{t+1} to be a proper distribution.
        t++ // incrementing t by 1
        end
end

Evaluation Metrics: Confusion matrix is used to analyze the performance and accuracy of any supervised learning algorithm. The confusion matrix will be used for the same.

TP = True Positive (correctly identified).
TN = True Negative (incorrectly identified).
FP = False Positive (correctly rejected).
FN = False Negative (incorrectly rejected).

Using the confusion matrix, the performance of the algorithm/technique can be calculated easily.

$$\text{Accuracy(Acc)} = \frac{(TP + TN)}{(TP + TN + FP + FN)}$$

$$\text{Sensitivity(Sen)} = \frac{(TP)}{(TP + FN)}$$

$$\text{Specificity(Spec)} = \frac{(TN)}{(TN + FP)}$$

$$F1\,\text{Score(Sen)} = \frac{(2TP)}{(2TP + FP + FN)}$$

Matthews Correlation Coefficient (MCC)

$$= \frac{(TP * TN) - (FP \times FN)}{\sqrt{(TP+FP)(TP+FN)(TN+FP)(TN+FN)}}$$

16.24 CONCLUSIONS

This chapter offers a comprehensive discussion about the cyberthreats and energy thefts in the smart grids and smart meters. Various solutions proposed by different researchers are also discussed in this chapter. The focus remained primarily on cyberattacks and energy thefts on AMIs in electrical infrastructure. An in-depth discussion about cybersecurity, particularly on the examination of network vulnerabilities and energy thefts in smart meters, is done. Our aim was to provide a deep understanding of cybersecurity vulnerabilities in smart grids and the detection of energy theft using smart meters and to guide on the future research directions on these terms. Data need to be handled properly and all the compliance frameworks regarding security should be adopted and a quality security system should be put in place. For proper protection, companies must understand that apart from compliance, security needs to be on focus, which is also a big part of compliance. We know that the transformation of the power system from centralized to decentralized topology uses new technology on the existing infrastructure for efficient and reliable execution [48]. The features in blockchain make its application to smart grids very promising. In this chapter, significant security challenges of smart grid scenarios that can be addressed by using many methods are discussed. A number of new methods such as blockchain-based security method, IoT-based method, and other recent research works are summarized. Several related practical projects, trials, and products that have recently been used are mentioned. The security regarding the avoidance of energy theft is revealed by using considerable information and historical consumption patterns of the consumers. The areas where blockchain is suggested in smart grid are AMIs, decentralized energy trading and market, and smart grid resilience. A smart meter can also be connected in blockchain to send the record and create a new block having its timestamp in a distributed ledger for future verification. This verification will ensure foolproof energy billing, selling, and purchasing/consuming. In this way, the consumer is charged strictly based on the data recorded on the ledger. The use of IoT (Internet of things) can also give efficient security mechanisms as power grids are accessed as distributed and scalable energy resources. This large-scale integration with advancements in energy utilization, greater efficiency, and reliability and ensuring security can be provided by using IoT, too. Here, the application of IoT in smart grids (IoT is an emerging technology having numerous advantages such as support to heterogeneous network structures with wide application areas, ability to communicate among varied devices, and higher security features) provide security in environments such as smart cities, smart meters, and smart energy management infrastructures.

REFERENCES

1. Z. A. Khan, M. Adil, N. Javaid, M. N. Saqib, M. Shafiq, and J. G. Choi, "Electricity theft detection using supervised learning techniques on smart meter data," *Sustainability*, vol. 12, no. 19, pp. 1–25, 2020, doi:10.3390/su12198023.
2. V. C. Gungor, et al., "Smart grid technologies: Communication technologies and standards," *IEEE Trans. Ind. informatics*, vol. 7, no. 4, pp. 529–539, 2011.
3. G. M. D. T. Forecast, "Cisco visual networking index: Global mobile data traffic forecast update, 2017–2022," *Update*, vol. 2017, p. 2022, 2019.
4. E. Sisinni, A. Saifullah, S. Han, U. Jennehag, and M. Gidlund, "Industrial internet of things: Challenges, opportunities, and directions," *IEEE Trans. Ind. Informatics*, vol. 14, no. 11, pp. 4724–4734, 2018.
5. L. Kotut and L. A. Wahsheh, "Survey of cyber security challenges and solutions in smart grids," in *2016 Cybersecurity Symposium (CYBERSEC)*, Canada, 2016, pp. 32–37.
6. P. Jokar, N. Arianpoo, and V. C. M. Leung, "RESEARCH ARTICLE A survey on security issues in smart grids," no. June 2012, pp. 262–273, 2016, doi:10.1002/sec.
7. S. S. Ali and B. J. Choi, "State-of-the-art artificial intelligence techniques for distributed smart grids: A review," *Electronics*, vol. 9, no. 6, p. 1030, 2020, doi:10.3390/electronics9061030.
8. Z. Lu, X. Lu, W. Wang, and C. Wang, "Review and evaluation of security threats on the communication networks in the smart grid," in *2010-Milcom 2010 Military Communications Conference*, pp. 1830–1835, 2010. Held on 31 October - 3 November 2010, San Jose, California, USA.
9. L. M. Camarinha-Matos, "Collaborative smart grids–a survey on trends," *Renew. Sustain. Energy Rev.*, vol. 65, pp. 283–294, 2016.
10. S. M. Amin and B. F. Wollenberg, "Toward a smart grid: Power delivery for the 21st century," *IEEE power energy Mag.*, vol. 3, no. 5, pp. 34–41, 2005.
11. B. B. Gupta and T. Akhtar, "A survey on smart power grid: Frameworks, tools, security issues, and solutions," *Annals of Telecommunications*, vol. 72, no. 9, pp. 517–549, 2017.
12. M. Z. Gunduz and R. Das, "Analysis of cyber-attacks on smart grid applications," *2018 Int. Conf. Artif. Intell. Data Process.* IDAP 2018, pp. 1–5, 2019, doi:10.1109/IDAP.2018.8620728.
13. F. Halim, S. Yussof, and M. E. Rusli, "Cyber security issues in smart meter and their solutions," *Int. J. Comput. Sci. Netw. Secur.*, vol. 18, no. 3, pp. 99–109, 2018.
14. A. Sanjab, W. Saad, I. Guvenc, A. Sarwat, and S. Biswas, "Smart grid security: Threats, challenges, and solutions," *arXiv Prepr. arXiv1606.06992*, 2016.
15. M. Baykara and R. Das, "A novel honeypot based security approach for real-time intrusion detection and prevention systems," *J. Inf. Secur. Appl.*, vol. 41, pp. 103–116, 2018, doi:10.1016/j.jisa.2018.06.004.
16. B. S. In, "Smart-grid security issues," *IEEE Secur. Priv.*, vol. 8, no. February, pp. 81–85, 2010.

17. V. Bolgouras, C. Ntantogian, E. Panaousis, and C. Xenakis, "Distributed key management in microgrids," *IEEE Trans. Ind. Informatics*, vol. 16, no. 3, pp. 2125–2133, 2020, doi:10.1109/TII.2019.2941586.
18. A. Stefanov and C. C. Liu, "Cyber-power system security in a smart grid environment," *2012 IEEE PES Innov. Smart Grid Technol. ISGT 2012*, pp. 6–8, 2012, doi:10.1109/ISGT.2012.6175560.
19. W. Wang and Z. Lu, "Cyber security in the smart grid : Survey and challenges," *Comput. Networks*, vol. 57, no. 5, pp. 1344–1371, 2013, doi:10.1016/j.comnet.2012.12.017.
20. N. Komninos, E. Philippou, and A. Pitsillides, "Survey in smart grid and smart home security: Issues, challenges and countermeasures," *IEEE Commun. Surv. Tutorials*, vol. 16, no. 4, pp. 1933–1954, 2014.
21. A. O. Otuoze, M. W. Mustafa, and R. M. Larik, "Smart grids security challenges: Classification by sources of threats," *J. Electr. Syst. Inf. Technol.*, vol. 5, no. 3, pp. 468–483, 2018, doi:10.1016/j.jesit.2018.01.001.
22. C. Bekara, "Security issues and challenges for the IoT-based smart grid," in *FNC/MobiSPC*, 2014, pp. 532–537.
23. A. Procopiou and N. Komninos, "Current and future threats framework in smart grid domain," in *2015 IEEE International Conference on Cyber Technology in Automation, Control, and Intelligent Systems (CYBER)*, 2015, pp. 1852–1857.
24. J. Kamto, L. Qian, J. Fuller, and J. Attia, "Light-weight key distribution and management for Advanced Metering Infrastructure," *2011 IEEE GLOBECOM Work. GC Workshops 2011*, pp. 1216–1220, 2011, doi:10.1109/GLOCOMW.2011.6162375.
25. X. Li, I. Lille, and N. Europe, "Securing smart grid: Cyber attacks, countermeasures, and challenges," *IEEE Commun. Mag.*, vol. 50,, no. August, pp. 38–45, 2012.
26. B. Khelifa and S. Abla, "Security concerns in smart grids: Threats, vulnerabilities and countermeasures," *Proc. 2015 IEEE Int. Renew. Sustain. Energy Conf. IRSEC 2015*, 2016, doi:10.1109/IRSEC.2015.7454963.
27. S. Shapsough, F. Qatan, R. Aburukba, F. Aloul, and A. R. Al Ali, "Smart grid cyber security: Challenges and solutions," in *2015 International Conference on Smart Grid and Clean Energy Technologies (ICSGCE)*, Offenburg, Germany, 20-23 October 2015, pp. 170–175.
28. Y. Yan, Y. Qian, H. Sharif, and D. Tipper, "A survey on smart grid communication infrastructures: Motivations, requirements and challenges," *IEEE Commun. Surv. Tutorials*, vol. 15, no. 1, pp. 5–20, 2013, doi:10.1109/SURV.2012.021312.00034.
29. D.B. Rawat DB and C. Bajracharya, "Cyber security for smart grid systems: Status, challenges and perspectives," In *SoutheastCon*, IEEE, Fort Lauderdale, FL, USA, 1–6, 2015.
30. E. U. Haq, H. Xu, L. Pan, and M. I. Khattak, "Smart grid security: Threats and solutions," in *Proc. -2017 13th Int. Conf. Semant. Knowl. Grids, SKG 2017*, vol. 2018-January, pp. 188–193, 2017, doi:10.1109/SKG.2017.00039.
31. A. Muir and J. Lopatto, "Final report on the August 14, 2003 blackout in the United States and Canada: Causes and recommendations," *US–Canada Power Syst. Outage Task Force*, Canada, 2004.
32. Y. Zhang, L. Wang, W. Sun, R. C. G. Ii, and M. Alam, "Distributed intrusion detection system in a multi-layer network architecture of smart grids," *IEEE Trans. Smart Grid*, vol. 2, no. 4, pp. 796–808, 2011, doi:10.1109/TSG.2011.2159818.
33. R. Ma, H. Chen, Y. Huang, W. Meng, and S. Member, "Smart grid communication: Its challenges and opportunities," *IEEE Trans. Smart Grid*, vol. 4, no. 1, pp. 36–46, 2013.
34. Y. Zhang, et al., "Wide-area frequency monitoring network (FNET) architecture and applications," *IEEE Trans. Smart Grid*, vol. 1, no. 2, pp. 159–167, 2010, doi:10.1109/TSG.2010.2050345.
35. V. C. Gungor and F. C. Lambert, "A survey on communication networks for electric system automation," *Comput. Networks*, vol. 50, no. 7, pp. 877–897, 2006.

36. T. Liu, Y. Sun, Y. Liu, Y. Gui, Y. Zhao, and C. Shen, "Abnormal traffic-indexed state estimation: A cyber-physical fusion approach for smart grid attack detection," *Futur. Gener. Comput. Syst.*, 2014, doi:10.1016/j.future.2014.10.002.

37. G. Garner, *"Designing Last Mile Communications Infrastructures for Intelligent Utility Networks (Smart Grids),"* IBM Australia Limited, 2010.

38. R. K. Pandey and S. M. Ieee, "Cyber security threats - Smart grid infrastructure," in *2016 National Power Systems Conference (NPSC)*, 2016, pp. 1–6.

39. S. De Dutta and R. Prasad, "Security for smart grid in 5g and beyond networks," *Wirel. Pers. Commun.*, vol. 106, no. 1, pp. 261–273, 2019, doi:10.1007/s11277-019-06274-5.

40. V. Galetić, I. Bojić, M. Kušek, G. Ježić, S. Dešić, and D. Huljenić, "Basic principles of Machine-to-Machine communication and its impact on telecommunications industry," in *2011 Proceedings of the 34th International Convention MIPRO*, Opatija, Croatia, 2011, pp. 380–385.

41. K. P. Soman and M. Alazab, "A comprehensive tutorial and survey of applications of deep learning for cyber security," *TechRxiv*, no. Dl, pp. 1–82, 2020, doi:10.36227/techrxiv.11473377.v1.

42. X. G. Shan and J. Zhuang, "A game-theoretic approach to modeling attacks and defenses of smart grids at three levels," *Reliab. Eng. Syst. Saf.*, vol. 195, p. 106683, 2020, doi:10.1016/j.ress.2019.106683.

43. W. Kong, J. Shen, P. Vijayakumar, Y. Cho, and V. Chang, "A practical group blind signature scheme for privacy protection in smart grid," *J. Parallel Distrib. Comput.*, vol. 136, pp. 29–39, 2020, doi:10.1016/j.jpdc.2019.09.016.

44. R. Marah, I. El Gabassi, S. Larioui, and H. Yatimi, "Security of smart grid management of smart meter protection," *2020 1st Int. Conf. Innov. Res. Appl. Sci. Eng. Technol. IRASET 2020*, 2020, doi:10.1109/IRASET48871.2020.9092048.

45. P. Zhuang, T. Zamir, and H. Liang, "Blockchain for cyber security in smart grid: A comprehensive survey," *IEEE Trans. Ind. Informatics*, vol. 3203, no. c, pp. 1–1, 2020, doi:10.1109/tii.2020.2998479.

46. A. Xiang and J. Zheng, "A situation-aware scheme for efficient device authentication in smart grid-enabled home area networks," *Electronics*, vol. 9, no. 6, p. 989, 2020, doi:10.3390/electronics9060989.

47. M. F. Moghadam, M. Nikooghadam, A. H. Mohajerzadeh, and B. Movali, "A lightweight key management protocol for secure communication in smart grids," *Electr. Power Syst. Res.*, vol. 178, no. April 2019, p. 106024, 2020, doi:10.1016/j.epsr.2019.106024.

48. M. B. Mollah et al., "Blockchain for future smart grid: A comprehensive survey," *IEEE Internet Things J.*, vol. X, no. vi, pp. 1–1, 2020, doi:10.1109/jiot.2020.2993601.

Index

Printed in the United States
by Baker & Taylor Publisher Services